D0072179

WITHDRAWN

Food Demand Analysis

PROBLEMS, ISSUES, AND EMPIRICAL EVIDENCE

Food Demand Analysis

PROBLEMS, ISSUES, AND EMPIRICAL EVIDENCE

EDITED BY
Robert Raunikar
AND
Chung-Liang Huang

 Iowa State University Press / Ames

Robert Raunikar is professor of agricultural economics, University of Georgia's Agricultural Experiment Station at Griffin.

Chung-Liang Huang is associate professor of agricultural economics, University of Georgia's Agricultural Experiment Station at Griffin.

Initial phase of editing by **Robert E. Branson** (Part I), **Oral Capps, Jr.** (Part II), **Daniel S. Tilley** (Part III), and **Karen J. Morgan** and **Chung-Liang Huang** (Part IV).

Printed in the United States of America from camera-ready copy provided by the editors

First edition, 1987

Library of Congress Cataloging-in-Publication Data

Food demand analysis.

    Includes bibliographies.
    1. Food consumption—Econometric models. I. Raunikar, Robert. II. Huang, Chung-liang, 1945– .
HD9000.5.F5947   1987   338.4′76413   86–30534
ISBN 0–8138–1841–9

Contents

Preface, vii

Introduction, ix

Contributors, xv

Part I.   THEORY AND DATA FOR DEMAND ANALYSES

1. Concepts of Consumer Demand Theory, 3

2. Data Sources for Demand Analyses, 33

3. Data Problems in Demand Analyses: Two Examples, 54

Part II.   COMPLETE DEMAND SYSTEMS

4. Comparison of Estimates from Three Linear Expenditure Systems, 91

5. Persistence in Consumption Patterns: Alternative Approaches and an Application of the Linear Expenditure System, 114

6. Analysis of Household Demand for Meat, Poultry, and Seafood Using the S_1-Branch System, 128

7. Analysis of Food and Other Expenditures Using a Linear Logit Model, 143

8. Complete Demand Systems and Policy Analysis, 154

Part III.   PARTIAL SYSTEMS: FACTORS INFLUENCING FOOD PURCHASES

    9.   Partial Systems of Demand Equations with a Commodity
         Emphasis, 171

    10. Socioeconomic, Demographic, and Psychological
         Variables in Demand Analyses, 186

Part IV.   POLICY ISSUES AFFECTING NUTRITION

    11. Consumer Demand for Nutrients in Food, 219

    12. Food Consumption and Nutrient Intake Patterns of
         School-Age Children, 236

    13. Impact of the Food Stamp Program on Food Expenditures
         and Diet, 255

Index, 281

Preface

IN SEPTEMBER 1964, the Department of Agricultural Economics at the Georgia Experiment Station proposed to the agricultural economics department heads at land grant institutions in the Southern Region that a Southern Regional Consumer Demand Project be initiated. The proposal was reviewed by selected persons in the Southern Region and subsequently submitted to heads of agricultural economics departments to obtain an indication of interest. As a result, a meeting of state representatives was held in July 1965 to develop a final project statement. The initial regional project, entitled "The Demand for Food" and designated SM-34, began in January 1966 with representation from Arkansas, Georgia, North Carolina, South Carolina, Tennessee, and USDA. The SM-34 project was followed by the SM-45 project for a five-and-one-half year period ending in December 1976 with representation from Georgia, Kentucky, North Carolina, Tennessee, Texas, and USDA.

The regional project, designated S-119, from which this book was developed included official technical committee representation from California, Florida, Georgia, Kentucky, Michigan, Minnesota, New York, North Carolina, Puerto Rico, Rhode Island, Tennessee, Texas, Virginia, Washington, Wisconsin, and USDA. Other participants in technical committee functions were from Canada and Missouri. This book introduces students and academicians to theoretical and applied literature on demand analysis generated from the technical committee. A wide spectrum of topics includes subject matter on food demand analysis recently addressed by the profession. Furthermore, it presents important additions to demand analysis and reveals needed areas of further investigations.

The authorship of this book is shared by numerous people who were either directly or indirectly involved in the regional research project. A listing of the contributing authors is provided.

In addition to authorship, contributions by others deserve an expression of gratitude. Robert E. Branson, Oral Capps, Chung-Liang Huang, Karen J. Morgan, and Daniel S. Tilley provided supervision of organization and editing of chapters within the four parts. The faculty and staff of the Department of Agricultural Economics at the Georgia Experiment Station provided a very understanding and accommodating spirit. A special debt of gratitude is extended to Nan Moon who typed the initial drafts of the book. Appreciation is also expressed to Sandra Danielly, Margie Lusk and Wadra McCullough for typing several drafts of the manuscript and the camera-ready copy. Finally, a sincere thanks to our colleagues on the S-119 project for their contributions to this book.

Introduction

"Some . . . (seem) to be mainly concerned with the statistical application of the theory, rather than with the theory itself." (Hicks 1956)

"The important thing is to coordinate the statistical methods with a thorough knowledge of the field under analysis, making use of all sorts of experience and prior information." (Wold and Jureen 1953)

THE ABILITY TO DESCRIBE ECONOMIC PHENOMENON has evolved through the aid of statistical methods which provide greater definition to economic relationships. However, in the absence of sound economic theory, experience, and prior information, interpretations that appear on the surface to be sound may, in fact, be quite hollow. What follows in the four parts of this book are discussions of economic theory and analytical procedures intended to expand the knowledge frontier of food demand and consumption behavior.

The study of food demand and consumption behavior has been restricted in the past by lack of data bases and lack of emphasis on the contribution which research results could provide for food policy. However, the recent availability of useful data bases and the accompanying increased interest in demand analysis have contributed to broadening the scope of research in the area. This has resulted in the opportunity to review the underlying economic theory of demand and to address some of the problems and issues associated with demand analysis. This volume brings together both a review of and current contributions to demand analysis. The discussion starts with basic information on economic theory and data. This is followed by a survey of recent applications and interpretations of both complete and partial demand systems. Finally, nutritional adequacy of food as it relates to food purchase behavior and the effects of public policy on nutrition are discussed.

Part I deals with theory and data for demand analyses and comprises three chapters. These chapters provide some of the basic framework from which demand analyses proceed. Additional contributions to theory and data complete the exposition required for topics presented in chapters which follow.

Chapter 1 by Capps and Havlicek reviews several important contributions to consumer demand theory. To develop the theoretical framework for consumer demand analysis, five topics are discussed: (1) concepts of neoclassical theory, (2) the derivation and analysis of several major properties of demand functions, (3) analysis of the structure of preferences, (4) extensions of neoclassical theory, and (5) key issues in demand analysis.

In Chapter 2, Branson and others provide an overview of the sources of data used in demand analyses. The first section of this chapter introduces basic concepts in public and private sources of data, data accuracy, the importance of data time frames, differentiation between aggregative and cross-sectional or market data, the differing levels in the market channel of data and data bases, and the needed data source and content definition. The second section describes selected data sources currently or potentially useful for demand analyses. The sources discussed include disappearance data systems, household panels, store panels, household surveys, channel movement data, economic censuses, and new data systems.

Chapter 3 provides some insights into and examples of data problems that arise in demand analyses. In the first example, Bobst relates the problems of the extent of measurement errors, the magnitudes of measurement error, and effect of measurement errors on demand estimation when disappearance data systems are used. In the second example, Buse cites problems, uses, and abuses in the application of consumer expenditure surveys to demand analyses. Specifically discussed are the conflicts between data and model, the conflicts between model selection and the resulting parameter estimates, and the importance of and improvement needed in public use data bases.

Part II consists of five chapters on the use of complete demand systems for analyzing consumer behavior with regard to food and nonfood items. In each chapter the emphasis is on the application and use of the particular complete demand systems formulations. At the beginning of each chapter, the particular model used is discussed briefly.

In Chapter 4, Craven and Haidacher use Stone's linear expenditure system, Leser's approximation, and Powell's approximation to analyze the demand for eleven expenditure categories including aggregate food consumed at home, aggregate food consumed away from home, and nonfoods. Marginal budget shares, total expenditure elasticities, and uncompensated price elasticities are estimated and analyzed for the eleven expenditure categories for the three alternative linear expenditure systems.

In Chapter 5, Green, Hassan, and Johnson use a dynamic system of demand equations which introduces a linear habit formation scheme into the linear expenditure system to analyze the demand for durables and semidurables, nondurables, and services in Canada. Six models entailing alternative specifications of the linear expenditure system and considering both the presence and absence of a first-order autoregressive error process are estimated and analyzed. Results from these models provide a basis for the hypothesis tests on habit persistence, serial correlation, and the classical consumer demand formulation. The empirical results demonstrate that various estimates are not robust to the different stochastic specifications and that models which employ alternative habit formation schemes should be tested relative to more generalized stochastic specifications.

In Chapter 6, Capps and Havlicek use the S_1-branch system to analyze the demand for meat, poultry, and seafood. In addition to the traditional economic variables of price and expenditure, sociodemographic variates are included in the complete demand systems model. The equations within the S_1-branch system are formulated in terms of quantities rather than expenditures. Own-price elasticities, expenditure elasticities, and system parameters are analyzed for eleven food and nonfood commodities for the United States and four component regions.

In Chapter 7, Tyrrell and Mount use a linear logit model to analyze food and other expenditures. This model entails choosing a flexible form of budget share equations with emphasis on characterizing the differences between households according to household size, household composition, and other household attributes. The linear logit model is used to analyze the effects of household income size and age characteristics on food budgets.

In the final chapter of this part, Chapter 8, Eastwood and Sun outline a framework and present a procedure for using elasticity and flexibility matrices generated by complete demand systems to evaluate alternative policy options. Guides for choosing an appropriate complete demand system and key things to consider in adapting demand models to answer policy questions are discussed. The framework developed in the beginning of the chapter is then used to analyze the retail market impacts of purchases of surplus food commodities by the U.S. Department of Agriculture under Section 32 of the Agricultural Adjustment Act.

Part III consists of two chapters which address the use of partial demand systems. The conditions under which partial demand systems have been used in demand analysis and how these systems are utilized are discussed.

In Chapter 9, Huang and Tilley with contributions by Bobst and Raunikar provide a discussion of partial demand systems on an individual commodity basis to provide for a better understanding of the unique nature of each analysis. The commodity analysis discussions concern (1) measurement of income-expenditure relationships of fresh whole milk and lowfat milk, (2) the analysis of demand for orange juice stressing the dynamic nature of two product forms, (3) examination of income-expenditure relationships across income levels using household beef expenditures, (4) estimation of price and income elasticities of regional demand for broiler meat, and (5) the use of a disequilibrium model for beef demand.

Chapter 10 focuses on the role of socioeconomic and psychological characteristics in explaining food expenditures and consumption. In the introduction, Chavas discusses the role of socioeconomic and demographic variables in both theoretical and applied research of demand analysis. The research represented in this chapter examines the role of socioeconomic variables within the context of utility theory, introduces psychological variables into food expenditure analysis, and presents extensive empirical results based on several sets of cross-sectional data. The empirical results are reported for four research efforts. In the work reported by Buse, attention is given to the effects of household characteristics on food expenditures. Specific emphasis is given to socioeconomic profiles, adult equivalent scales, and interaction effects among socioeconomic characteristics. In the next section, Price, West, and Price examine the effects of household age/sex composition on food consumption where 8- to 12-year-old children in the household are the units of observation. This research investigates the effect of basic need levels on types of food consumed and food expenditure levels. In the next body of research, Chavas proposes a more generalized approach to the Prais-Houthakker model. In the final empirical contribution to this chapter, Blaylock and Smallwood use the Lorenz and expenditure concentration curves to generate income elasticities. The chapter concludes with some generalization based on these empirical results and the identification of areas which deserve additional research efforts.

Parts II and III focus on obtaining a better understanding of food demand and consumption behavior of U.S. households under complete and partial demand system approaches, respectively. In recent years, the nutritional aspect of American food habits has been of considerable interest to both policy-makers and the general public. In Part IV, the importance of dietary components of foods as related to consumer food purchasing behavior and the effects of public policy on dietary quality and adequacy for some specific target groups of consumers are addressed and investigated.

In Chapter 11, the first chapter in Part IV, Morgan uses the hedonic index technique to investigate the relationships between price per ounce of breakfast cereals and their dietary components. The hypothesis that dietary components of breakfast cereals are relevant attributes which affect consumer's decision in purchasing breakfast cereals is tested.

In Chapter 12, specific issues concerning the nutritional status and nutrient intake patterns of school-age children are addressed. Morgan reports first on the effects of consumption of ready-to-eat cereal and salted snack foods on nutrient intake of 5- to 12-year-old school children. In addition, the sources of sugar in children's diets are also investigated. Price reports in the second section of Chapter 12 concerning the effects of the school lunch program and the food stamp program (FSP) on the nutrient intake and food consumption patterns of 8- to 12-year-old school children in the state of Washington.

Chapter 13 consists of four sections concerned with various issues related to the FSP. In the first section, Senauer identifies and analyzes the factors that may help explain FSP participation among eligible households. In the second section, Schrimper uses two alternative approaches to examine the effects of the FSP on food expenditures. One approach is based on comparison of individual household food expenditure patterns. The second approach is based on aggregate analyses to identify the impact of the FSP on the total demand for food and its subsequent effects on food prices. The last two sections of Chapter 13 examine the critical issue of whether or not the FSP increases nutrient intake of the participants. Based on the data collected from the state of Washington, Price reports that among 8- to 12-year-old school children, intake of nutrients as a result of participation in the FSP is, in general, not statistically significant. However, there is some indication of improvement in the nutrient quality of the diet with FSP participation. Using nutrients derived from foods purchased by low-income households in the South, Huang reports that the FSP significantly increases the amount of food nutrients available to participants except for those households whose head is 65 years old or older.

REFERENCES

Hicks, J. R. 1956. _A_ _revision_ _of_ _demand_ _theory_. London: Oxford Univ. Press.
Wold, H., and L. Jureen. 1953. _Demand_ _analysis_. New York: Wiley.

Contributors

James Blaylock, Economist, Economic Research Service, U.S. Department of Agriculture.

Barry W. Bobst, Associate Professor, Department of Agricultural Economics, University of Kentucky.

Robert E. Branson, Professor, Department of Agricultural Economics, Texas A and M University.

Reuben C. Buse, Professor, Department of Resource Economics, University of Wisconsin.

Oral Capps, Jr., Associate Professor, Department of Agricultural Economics, Texas A and M University. Formerly Associate Professor, Departments of Agricultural Economics and Statistics, Virginia Polytechnic Institute and State University.

Jean-Paul Chavas, Associate Professor, Department of Resource Economics, University of Wisconsin. Formerly Assistant Professor, Department of Agricultural Economics, Texas A and M University.

John A. Craven, Economist, Economic Research Service, U.S. Department of Agriculture.

David B. Eastwood, Associate Professor, Department of Agricultural Economics and Rural Sociology, University of Tennessee.

Richard Green, Associate Professor, Department of Agricultural Economics, University of California at Davis.

Richard C. Haidacher, Economist, Economic Research Service, U.S. Department of Agriculture.

Zuhair Hassan, Economist, Agriculture Canada.

Joseph Havlicek, Jr., Professor and Chair, Department of Agricultural Economics, Ohio State University. Formerly Professor, Departments of Agricultural Economics and Statistics, Virginia Polytechnic Institute and State University.

Chung-Liang Huang, Associate Professor, Department of Agricultural Economics, University of Georgia.

Eva E. Jacobs, Statistician, Bureau of Labor Statistics, U.S. Department of Labor.

Stanley R. Johnson, Professor of Economics and Administrator, Center for Agricultural and Rural Development, Iowa State University. Formerly Professor, Department of Agricultural Economics, University of Missouri-Columbia.

Karen J. Morgan, Associate Professor, Department of Human Nutrition, Foods and Food Systems Management, University of Missouri-Columbia.

Timothy D. Mount, Professor, Department of Agricultural Economics, Cornell University.

David W. Price, Professor, Department of Agricultural Economics, Washington State University.

Dorothy Z. Price, Professor, Home Economics Research Center, Washington State University.

Robert Raunikar, Professor, Department of Agricultural Economics, University of Georgia.

Ronald A. Schrimper, Professor, Department of Economics and Business, North Carolina State University.

Benjamin J. Senauer, Professor, Department of Agricultural and Applied Economics, University of Minnesota.

David Smallwood, Economist, Economic Research Service, U.S. Department of Agriculture.

Theresa Y. Sun, Economist, Economic Research Service, U.S. Department of Agriculture.

Daniel S. Tilley, Associate Professor, Department of Agricultural Economics, Oklahoma State University. Formerly Associate Professor, Food and Resource Economics Department, University of Florida.

Timothy J. Tyrrell, Associate Professor, Department of Resource Economics, University of Rhode Island.

Donald A. West, Economist, Extension Service, U.S. Department of Agriculture.

Lillian R. de Zapata, Economist, Department of Economics, Puerto Rico Agricultural Experiment Station.

PART I

Theory and Data
for Demand Analyses

CHAPTER 1

Concepts of Consumer Demand Theory

"But it is an equal empirical truth that the facts do not tell their own story to scientists or historical observers, and that the men who develop top-notch judgement have an analytical framework within which they try to fit the facts." (Samuelson, 1965)

THE NINETEENTH CENTURY marks the inception of the theory and measurement of consumer behavior principally through the works of Cournot, Dupuit, Marshall, Walras, Gossen, Jevons, Edgeworth, Antonelli, Engel, and Fisher. Since then, the theory and measurement of consumer behavior have developed enormously, as typified by the monumental monographs of Slutsky, Allen and Bowley, Schultz, Hicks, Wold and Jureen, Stone, Brandow, and George and King. Demand analysis relies on the integration of economic theory and statistical methods.

The purpose of this chapter is to present the theoretical framework which lies behind the empirical work on consumer demand, thereby laying the foundation for the remaining chapters of this monograph. The discussion focuses on the following topics: (1) concepts of neoclassical theory, (2) the derivation and analysis of several major properties of demand functions, (3) analysis of the structure of preferences, (4) extensions of neoclassical theory, (5) key issues in demand analysis, and (6) summary.

CONCEPTS OF NEOCLASSICAL THEORY

Neoclassical demand theory attempts to explain how individuals make consumption decisions at a given point in time.

Authors of this chapter are Oral Capps, Jr., and Joseph Havlicek, Jr.

The building blocks of the theory include the utility function, the commodity set, and the axioms concerning the ordering of preferences.

The Utility Function

As a starting point in the theoretical development of consumer demand, one must assume the existence of a scalar, continuous utility function, U, and in addition, assume the existence of a finite number of commodities, say n $(q_1, \ldots, q_n)$.

The utility function $U = U(q_1, \ldots, q_n)$ is an interpretive measure of the satisfaction derived from the consumption of alternative commodity bundles; U is a number or scalar that facilitates the ordering of the commodity bundles according to the consumer's preferences. Thus, the utility function is a numerical representation of a preference ordering. Particularly noteworthy is the following fundamental assumption, the postulate of rationality: the consumer chooses among the available alternative bundles with the objective of maximizing personal satisfaction. This rationality postulate is the customary departure point in consumer behavior theory.

The Commodity Set

Three properties are used to define the commodity set: (1) the nonnegativity property, (2) the divisibility property, and (3) the unboundedness property. The first property states that no commodity bundle may have negative components. In conjunction with the second property, let $q^0 = (q_1^0, \ldots, q_n^0)$ be a bundle available to the consumer. Then any bundle of the form $\alpha q^0 = (\alpha q_1^0, \ldots, \alpha q_n^0)$, for $0 \leq \alpha \leq 1$, may be extracted from this bundle. The third property states that the commodity set contains the bundle $(0, \ldots, 0)$, and moreover, if a bundle q^1 belongs to the set, then any bundle q^2, where $q_i^2 \geq q_i^1$ for all i belongs to the set, that is, the set is unbounded from above.

Preference Axioms

Several preference axioms (comparability, antisymmetry, transitivity, continuity, monotonicity, convexity, and differentiability) are necessary and sufficient conditions to insure the existence of order-preserving, monotonic, quasi-concave, real-valued continuous utility functions (Debreu 1959).[1] These axioms plus the utility function concept and the commodity set concept have consequences for applied work. In particular, a given preference ordering defines a set of demand relationships. When preferences change, this set also changes since the form of the utility function determines the form of the demand relationships. The implicit assumption in any empirical analysis is that the utility function does not change over the observation period.

DERIVATION AND ANALYSIS OF PROPERTIES
OF CONSUMER DEMAND FUNCTIONS
    The neoclassical demand model forms the starting point for the treatment of consumer behavior. Certain properties evolve from the derivation and analysis of the set of demand equations which result from utility maximization. These properties take the form of mathematical restrictions on the derivatives of the demand functions, which are always effective regardless of the form of the utility function.

Derivation of the System of Demand Functions
    Assume that the consumer has a given income (or total expenditure), say y, and in addition, assume that the consumer faces a choice among alternative commodity bundles from the commodity space. Let $p = (p_1, \ldots, p_n)$ represent the prices per standard unit of the commodities in the various commodity bundles. The choice problem for the consumer is then summarized by the following. Given p and y, the consumer selects the bundle to maximize utility $U = U(q)$, where $q = (q_1, q_2, \ldots, q_n)$, subject to the income or budget constraint $\Sigma_{i=1}^{n} p_i q_i = y$.

    The consumer's choice of individual commodities, $q_1, \ldots, q_n$, corresponds to the quantities consistent with the maximization of

$$L(q,\lambda) = U(q_1, \ldots, q_n) - \lambda(p_1 q_1 + \ldots + p_n q_n - y) \qquad (1.1)$$

where λ, a scalar, is the Lagrangian multiplier that is interpreted as the marginal utility of income. Differentiation with respect to $q_1, \ldots, q_n$ and λ gives

$$U_i(q_1, \ldots, q_n) - \lambda p_i \qquad (i = 1, \ldots, n) \qquad (1.2)$$

and $y - p_1 q_1 - \ldots - p_n q_n$, where $U_i = \partial U/\partial q_i$ is the marginal utility of the ith commodity.

    The axiom of monotonicity insures that $U_i > 0$ for all i; the marginal utilities are everywhere positive. When these derivatives are set equal to zero,

$$U_i = \lambda p_i \quad \text{and} \quad y = \sum_{i=1}^{n} p_i q_i \qquad (i = 1, \ldots, n) \qquad (1.3)$$

The result is the familiar first-order conditions of utility maximization. The system of relationships in (1.3) consists of n + 1 equations in the n + 1 unknowns of $q_1, \ldots, q_n$ and λ when p and y are given. Hence, the first-order conditions may be solved for the quantities that provide the individual with the highest possible level of satisfaction. The solutions to the n + 1 normal equations may be expressed as

$$q_i = q_i(p_1, \ldots, p_n, y) \quad \text{and} \quad \lambda = \lambda(p_1, \ldots, p_n, y) \qquad (1.4)$$

The quantity purchased of each commodity and the marginal utility of income are expressed as a function of the prices of the respective commodities and income. The n equations $q_i = q_i(p_1, \ldots, p_n, y)$ constitute the complete set of consumer demand functions. Obtaining this set of (Marshallian) demand functions is the main objective of classical consumer demand theory.

The relationships in (1.3) only assure that the consumer is at a stationary or saddle point on the Lagrangian function. This stationary point will correspond to either a maximum or a minimum point of satisfaction. Define the Hessian matrix V by

$$V = \begin{bmatrix} U_{11} & \cdots & U_{1n} \\ \vdots & & \\ U_{n1} & \cdots & U_{nn} \end{bmatrix} \qquad (1.5)$$

where $U_{ij} = U_{ji} = \partial^2 U / \partial q_i \partial q_j = \partial U_i / \partial q_j$. Symmetric matrix V is negative definite, due to the axioms of monotonicity, convexity, and differentiability, which ensures that the constrained maximization problem has a unique solution for q as a function of y and p. Consequently, the set of demand relationships satisfies the second-order conditions of utility maximization. Incidentally, this set of conditions implies that $U_{ii} < 0$ for all i so that diminishing marginal utility exists for each commodity.

Properties of the System of Demand Equations

A system of demand functions should also satisfy a number of crucial relationships or restrictions. Those comprising the main target of the classical theory are the Engel aggregation condition, the Cournot aggregation conditions, the homogeneity conditions, and the Slutsky symmetry conditions. Each of these conditions defines an exact set of relationships which any complete set of demand functions must possess, if it is derivable from the maximization of any utility function. Such relationships follow from the differentiation of the first-order conditions of utility maximization. A full mathematical treatment appears in several works (Goldberger 1967; Phlips 1974; Theil 1975; Barten 1977; and Deaton and Muellbauer 1980b).

The Engel aggregation (adding-up) condition deals with the

differentiation of the budget constraint with respect to income. This condition requires that the sum of the income elasticities [$n_i = (\partial q_i/\partial y)(y/q_i)$, $i = 1, \ldots, n$] weighted by their respective average budget shares ($w_i = p_i q_i/y$, $i = 1, \ldots, n$) equals one.

The Cournot aggregation conditions concern the effects of a change in the price of the jth commodity with all other ($n - 1$) prices invariant. In terms of uncompensated price changes, this set of conditions states that the sum of the cross-price elasticities [$n_{ij} = (\partial q_i/\partial p_j)(p_j/q_i)$, $i \neq j$, $i, j = 1, \ldots, n$] of the jth commodity plus the direct price elasticity [$n_{jj} = (\partial q_j/\partial p_j)(p_j/q_j)$, $j = 1, \ldots, n$] of the jth commodity, all weighted by their respective average budget shares, equals the negative of the average budget share for the jth commodity.

The homogeneity conditions imply that the demand functions are homogeneous of degree zero in prices and income. If all prices and income change proportionately, no changes in consumption levels occur. In terms of uncompensated price changes, this set of conditions states that the sum of the direct and cross-price elasticities plus the income elasticity for any commodity is zero.

The Slutsky symmetry conditions indicate the relationship among cross-price elasticities, average budget proportions, and income elasticities for any pairs of goods. The classical demand restrictions, including symmetry, expressed in elasticity form are shown in Table 1.1.

TABLE 1.1. Classical demand restrictions expressed in elasticity form in terms of uncompensated price changes

Classical restrictions	Mathematical representations
Engel aggregation condition	$\sum_{i=1}^{n} w_i n_i = 1$
Cournot aggregation conditions	$\sum_{i=1}^{n} w_i n_{ij} = -w_j$, $\quad j = 1, \ldots, n$
Slutsky symmetry conditions	$w_i(n_{ij} + n_i w_j) = w_j(n_{ji} + n_j w_i)$, $\quad i, j = 1, \ldots, n$
Homogeneity conditions	$\sum_{j=1}^{n} n_{ij} + n_i = 0$, $\quad i = 1, \ldots, n$

The demand functions $q = q(y, p)$ may be inserted into the utility function $U = U(q)$ to give maximum attainable utility as a function of y and p. Thus $U = U[q(y, p)] = \Phi(y, p)$. The function $\Phi(y, p)$ is the indirect utility function as distinguished from the direct utility function $U = U(q)$. Indirect utility functions express utility in terms of prices and income rather than quantities consumed directly. A relationship exists between the direct and indirect utility function since both represent the same preference ordering. Once the indirect utility function is specified, the demand functions are obtained by applying Roy's identity: $q_i = -(\partial\Phi/\partial p_i)/(\partial\Phi/\partial y)$. While the direct utility function probably has greater intuitive appeal, the indirect utility function is not without its claims to interest. In particular, inverting $U = \Phi(y, p)$ leads to the cost function $y = C(U, p)$, the minimum cost of attaining U at p. Therefore, the indirect utility function and the cost function lay the foundation of constant-utility index numbers useful for welfare analysis. Additionally, differentiating $C(U, p)$ with respect to p (Shephard's Lemma) leads to the Hicksian demand functions, $q_i = h_i(U, p)$. In fact, the indirect utility function and the cost function are at the heart of the duality relationships in problems of consumer choice (Deaton and Muellbauer 1980b).

This section relies on the indirect utility function to investigate further the analytic properties of a set of demand equations. Following Goldberger (1967), setting the total differential of $U = \Phi(y, p)$ to zero gives the income change which compensates for a price change in the sense of keeping utility unchanged. For infinitesimal changes, compensation that holds utility constant (Hicks) and compensation that allows the purchase of the same individual commodities after the price change (Slutsky) are equivalent. Alternatively, setting the total differential of $\lambda = \lambda(y, p)$ to zero gives the income change which compensates for a price change in the sense of keeping the marginal utility of income unchanged.

With these two concepts of compensation in hand, two decompositions of the effects of price changes upon quantities demanded are possible. The response to a Hicksian compensated or Slutsky compensated price change is given by

$$(\partial q_i/\partial p_j)^* = (\partial q_i/\partial p_j) + q_j(\partial q_i/\partial y)$$

$$= \lambda v^{ij} - \phi y(\partial q_i/\partial y)(\partial q_j/\partial y) \qquad (1.6)$$

This expression measures the response of the quantity demanded of the ith good to a change in the price of the jth good when that price change is accompanied by an income change which maintains the same level of utility. Note that $(\partial q_i/\partial p_j)$ is the uncompensated change of the quantity demanded of the ith good attributable to a change in the price of the jth good, $(\partial q_i/\partial p_j)^*$ is the compensated change, $(\partial q_i/\partial y)$ is the change in the quantity demanded of the ith good attributable to a change in income, v^{ij} denotes the ijth element of v^{-1}, the inverse of the Hessian matrix, and ϕ is the reciprocal of Frisch's money flexibility--the income elasticity of the marginal utility of income, $w = (\partial\lambda/\partial y)(y/\lambda)$.

Here w serves as an indicator of welfare, and equivalently,

ϕ serves as an inverse indicator of welfare. The presumption is that the marginal utility of income, λ, is very high at low-income levels and approaches zero at high-income levels with respect to consumable goods. Under this presumption, w moves from large negative values at low-income levels to small negative values at high-income levels.

Consequently, the total effect of an uncompensated price change $(\partial q_i / \partial p_j)$ is decomposed into a substitution effect $(\partial q_i / \partial p_j)^*$ and an income effect $q_j (\partial q_i / \partial y)$. Goods i and j are classified as substitutes, independent, or complements according as $(\partial q_i / \partial p_j)^* = (\partial q_j / \partial p_i)^*$ is positive, zero, or negative (Allen and Hicks 1934). The Slutsky symmetry conditions state that the compensated cross-price derivatives are symmetric. Economic theory has nothing to say about the sign of the income effect in the absence of a particular specification of the utility function.

The response to a marginal-utility-of-income-compensated price change is given by

$$(\partial q_i / \partial p_j)^{**} = \lambda v^{ij} = (\partial q_i / \partial p_j)^* + \phi y (\partial q_i / \partial y)(\partial q_j / \partial y) \quad (1.7)$$

This expression measures the response of the quantity demanded of the ith good attributable to a change in the price of the jth good when that price change is accompanied by an income change which just maintains the marginal utility of income. In this case, the total effect of an uncompensated price change has thus been decomposed into three terms: (1) a specific substitution effect, $(\partial q_i / \partial p_j)^{**}$; (2) a general substitution effect, $(\partial q_i / \partial p_j)^* - (\partial q_i / \partial p_j)^{**}$; and (3) an income effect, $q_j (\partial q_i / \partial y)$. Houthakker (1960) suggested that goods i and j might be reclassified as substitutes, independent, or complements according as $(\partial q_i / \partial p_j)^{**} = (\partial q_j / \partial p_i)^{**}$ is positive, zero, or negative, that is, according as $v^{ij} = v^{ji} \gtrless 0$.

Schematically, the following decomposition in algebraic form occurs which typically is labeled the Hicksian fundamental equation of value theory:

$$(\partial q_i / \partial p_j) = \lambda v^{ij} - \phi y (\partial q_i / \partial y)(\partial q_j / \partial y) - q_j (\partial q_i / \partial y) \quad (1.8)$$

Total Effect	Specific General Substitution Effect	Income Effect

This exposition shows that the substitution effect can be decomposed into two components. The first component, λv^{ij}, is called the specific substitution effect, and the second component, $-\phi y(\partial q_i/\partial y)(\partial q_j/\partial y)$, is the general substitution effect. When $i = j$, the decomposition refers to the direct substitution effect. When $i \neq j$, it refers to the cross-substitution effect. The adjectives "specific" and "general" are well chosen. The first indicates that the corresponding component depends upon the specific relation in terms of the ij elements of the inverse of the Hessian matrix V. In addition, since $(\partial q_i/\partial p_j)** = \lambda v^{ij}$, the specific substitution effect is also the response to a marginal-utility-of-income-compensated change. The second emphasizes that the corresponding component represents an overall effect. When no interactions at all exist among commodities, the V is a diagonal matrix. Interactions among commodities show up in the off-diagonal positions of the Hessian. Where V is diagonal, so is v^{-1}, and the cross-substitution effects reduce to $-\phi y(\partial q_i/\partial y)(\partial q_j/\partial y)$. When V is nondiagonal, the appropriate element of λv^{-1} adds together with $\phi y(\partial q_i/\partial y)(\partial q_j/\partial y)$ to obtain the substitution effect. Since the term λv^{-1} arises only through interaction among commodities, it is labeled the specific substitution effect.

Given the system of equations in (1.4) for n commodities, there exist n^2 price elasticities and n income elasticities, and therefore, a total of $n(n + 1)$ parameters needs to be estimated. Econometric techniques require the number of observations to equal or exceed the number of parameters. From the viewpoint of applied econometric analysis, the set of the classical restrictions discussed in this section serves to reduce the dimension of the parameter space. This set of relationships provides $(n^2 + n + 2)/2$ independent restrictions, and hence the dimension of the parameter space is reduced from $n(n + 1)$ to $(n^2 + n - 2)/2$.[2] If n, the number of goods, is sufficiently large, the number of observations can be larger than the number of parameters. This leads to the so-called degrees of freedom problem as indicated by Bieri and de Janvry (1972).

ANALYSIS OF THE STRUCTURE OF PREFERENCES

The problem of degrees of freedom that researchers often encountered in the empirical estimation of the system of demand equations may be resolved from a number of ad hoc approaches. Among the most common practices are: (1) using single commodity or sector models under the explicit assumption that omitted factors have zero effects; (2) increasing the number of available observations, for instance, by pooling time-series and cross-sectional data; (3) aggregating commodities with the use of the Leontief-Hicks composite commodity theorem--commodities whose relative prices do not change are in effect a single commodity for the purposes of the theory. These approaches are in effect directed to solve the problem by either increasing the number of observations or decreasing the number of commodities included in the system. Alternatively, one may approach the problem from a conceptual standpoint by imposing certain specific assumptions regarding the interaction of commodities and the nature of utility functions. This last approach forms the foundation of additive preferences, almost additive preferences, and separability conditions, which can be used to further reduce the dimension of the parameter space into manageable size.

Additive Preferences

The additive preferences or additivity assumption allows an account of the independence of certain aggregates or groups of commodities. A preference ordering is directly additive if there exists a differentiable function F, F' > 0, and n functions $U^i(q^i)$ such that $F[U(q_1, \ldots, q_n] = \Sigma_{i=1}^{n} U^i(q_i)$. Each $U^i(q_i)$ is a function of only the corresponding q_i. Synonyms of direct additivity include want-independence, complete want-independence, and preference independence. The venerable Klein-Rubin utility function, $U = \Sigma_{i=1}^{n} \beta \ln(q_1 - Y_i)$, is additive. Frisch (1959) and Houthakker (1960) demonstrated that direct additivity has important implications for a complete set of demand functions.

The key is the diagonality of the Hessian matrix V. If U(q) is directly additive, then the marginal utility of any good varies with the quantity of that good alone. So, $U_{ij} = 0$ for all i,j, such that i = j, and hence V as well as V^{-1} are diagonal. Specifically, $U^{ij} = U_{ii}^{-1}$ where i = j and $U^{ij} = 0$ otherwise. Some additional properties of interest emerge with adherence to direct additivity of the utility function. To illustrate, Houthakker showed that cross-price derivatives are proportional to derivatives with respect to income. Mathematically, this result is simply

$$(\partial q_i / \partial p_k) / (\partial q_j / \partial p_k) = (\partial q_i / \partial y) / (\partial q_j / \partial y) \quad (i \neq j \neq k) \quad (1.9)$$

Moreover, under direct additivity, a knowledge of the income flexibility (ϕ) or the income elasticity of the marginal utility of income (w) enables the computation of all price coefficients from the income coefficients. The possibility then exists for the estimation of price elasticities from cross-sectional samples without the observation of any price variation --a rather striking result.

In short, a directly additive utility function leads to

substantial economies in the parameterization of a complete set
of demand functions beyond those obtainable from the classical
theory.   Within this framework, only n parameters require
estimation. However, these economies are not costless. Direct
additivity rules out the possibility of specific substitution
effects, the possibility of inferior goods, and the possibility
of complementary goods. In light of these restrictions, direct
additivity may be a plausible specification when goods are broad
aggregates. For the case where a fine classification of goods
exists, this specification is not tenable. In the latter case,
specific substitution effects, complementary goods, and inferior
goods may be present.

Almost Additive Preferences

Barten (1964) assumed the possibility that consumers'
preferences are not completely additive. The definition of
almost additive preferences rests on the relative magnitudes of
the on- and off-diagonal elements of the Hessian matrix of the
utility function V.   The issue at stake is whether V is
"sufficiently diagonal" to allow a certain convenient decomposition of its inverse.

Let U_{ij} be the ijth element of V.   Then a necessary
condition for almost additive preferences to pertain is that
$|U_{ij}/(U_{ii}U_{jj})^{1/2}| < 1$ for all i, j(i $\neq$ j). If this result occurs, then a power series expansion for V^{-1} is available; write

$$V = -A(I + \bar{H}) A^{-1}$$
(1.10)

where A is an n x n diagonal matrix such that the iith element
on the diagonal is $(-U_{ii})^{1/2}$ ($U_{ii} < 0$ for a well-behaved utility

function). The matrix $\bar{H}$ is symmetric, has zeroes on the principal
diagonal, and the off-diagonal elements h^{ij} of $\bar{H}$ satisfy $-h_{ij}$
$(U_{ii}U_{jj})^{1/2} = U_{ij}$, i $\neq$ j. If $|U_{ij}/(U_{ii}U_{jj})^{1/2}| < 1$, then all of
the h_{ij} are less than one in absolute value, and the following
expansion occurs for $(I + \bar{H})^{-1}$:

$$(I + \bar{H})^{-1} = I - \bar{H} + \bar{H}^2 - \bar{H}^3 + \dots$$
(1.11)

Hence, $V^{-1} = -A^{-1}(I - \bar{H} + \bar{H}^2 - \bar{H}^3 + \dots)A^{-1}$ and by deletion of
higher-order terms,

$$V^{-1} \doteq -A^{-1}(I - \bar{H}) A^{-1}$$
(1.12)

If the approximation (1.12) is sufficiently adequate, then
almost additive preferences pertain.

Strictly speaking, however, this specification does not, without further restrictions, accomplish any parsimony of parameterization, for the elements of $\bar{H}$ are unknown. By specifying no interaction between certain pairs of commodities, a reduction in the number of parameters may occur. (An alternative could be assumed wherein all of the nondiagonal elements in $\bar{H}$ have the same value.) In this light, Barten (1964) assumed that V for the n good case is of the form:

$$
V = \begin{bmatrix} U_{1}{}_{k \times k} & \vdots & O_{k \times (n-k)} \\ \cdots\cdots\cdots\cdots\cdots\cdots\cdots\cdots\cdots\cdots\cdots\cdots \\ O_{(n-k) \times k} & \vdots & U_{2}{}_{(n-k) \times (n-k)} \end{bmatrix} \tag{1.13}
$$

where U_2 is a diagonal matrix of dimension $(n - k) \times (n - k)$. The matrix U_1 is a symmetric matrix such that

$$
U_1 = \begin{bmatrix} U_{11} - h_{12}(U_{11}U_{22})^{1/2} & \cdots & - h_{1k}(U_{11}U_{kk})^{1/2} \\ & U_{22} & \cdots & - h_{2k}(U_{22}U_{kk})^{1/2} \\ & & \ddots \\ & & & U_{kk} \end{bmatrix}
$$

$(h_{jk} < 1) \qquad (j \neq k)$ \hfill (1.14)

In particular, Barten (1964) allowed a small amount of interaction among commodities but the interaction is of such a nature that (1) V^{-1} is defined with the same off-diagonal zero elements as the V matrix, and (2) the nonzero off-diagonal elements have a relation to the geometric average of the corresponding diagonal elements in the same ways for both V and V^{-1}.

Let k_i be the number of commodities with which the ith commodity may interact in the utility function. For a given i, there exist k_i indices for which the first term on the right-hand side of equation (1.8), the specific substitution effect, does not vanish. For the almost additive preferences specification, the dimension of the parameter space diminishes due to $(n + 1) + (\Sigma_{i=1}^{n} k_i)/2$ independent constraints and the independence assumption among certain commodities. Seemingly, the almost additive preferences approach offers more heuristic appeal than the complete want-independence approach.

The Separability Approach
     The approaches by Frisch (1959), Houthakker (1960), and
Barten (1964) assume additivity and almost additivity of the
utility function, and hence, very little interaction among
commodities in the utility function. A concept of separability
evolves from Strotz's (1957) utility tree representation. There
has been a tendency in recent years for some economists to
consider restrictions in the form of various separability
conditions on the consumer's utility function.
     Separability is a relative concept whose frame of reference
is some partition of the complete set of n commodities into S
mutually exclusive and exhaustive subsets. In general, the
separability conditions require the marginal rates of substitution for certain pairs of commodities to be functionally independent of the quantities of certain other commodities. Restrictions such as these reduce the number of parameters that
enter into the family of demand functions and in short, make
estimation of the entire parameter space more feasible. In
general, three types of separability definitions are in exis-
tence--weak separability, strong separability, and Pearce separability. Each of these intuitively tenable relationships warrants considerations.

Weak separability. A utility function U is weakly separable
with respect to a partition of the commodity space if the marginal rate of substitution between any two goods i and j from
within the same subset, say G_s, is independent of the quantities of the commodities consumed from other subsets. Mathematically,

$$\partial(U_i/U_j)/\partial q_k = 0,$$

for all i, $j \epsilon G_s$, $k \epsilon G_s$,   s = 1, ..., S   (1.15)

Goldman and Uzawa (1964) showed that under this assumption the
utility function assumes a nonadditive form.

$$U(q_1, ..., q_n) = F[U^1(q^1), U^2(q^2), ..., U^S(q^S)]   (1.16)$$

where F is a scalar function of n variables, and $U^i(q^i) = U^i(q_1^i,$
$q_2^i, ..., q_{n_1}^i)$ (i = 1, 2, ..., S), the number of groups; $\Sigma_{i=1}^{S} n_i =$
n, is the total number of commodities. In addition, weak
separability is equivalent to Strotz's concept of a utility
tree, and further, this concept is a prerequisite for the
consistency of Strotz's two-stage maximization procedure.[3]
     Several implications arise from the weak separability condi-

tion. All cross-price elasticities between commodity groups require only the knowledge of the income elasticities and at least one intergroup coefficient. The intergroup coefficients are measures of the degree of substitutability among groups of goods. Further, the cross-price elasticities among commodity groups are proportional to the relevant income elasticities. These relationships suggest a substantial number of parameter restrictions which reduce the number of parameters to be estimated.

Strong separability. A utility function U is strongly separable with respect to the partition under consideration, if the marginal rate of substitution between any two commodities from different subsets does not depend upon the quantities of commodities not belonging to those subsets. In mathematical terms,

$$\partial(U_i/U_j)/\partial q_k = 0,$$

$$(i \; \varepsilon \; G_s, \qquad j \; \varepsilon \; G_r, \qquad k \; \varepsilon \; G_t, \qquad G_s \neq G_r \neq G_t,$$

$$s, \; r, \; t = 1, \; ..., \; S) \tag{1.17}$$

These conditions, according to Goldman and Uzawa (1964), imply a utility function which is additive among commodity groups:

$$U(q_1, \; ..., \; q_n) = F[U^1(q^1) + U^2(q^2) + \; ... \; + U^S(q^S)] \tag{1.18}$$

As a result of strong separability, the marginal utility of a commodity in one group is independent of the consumption of any good in any other group. Hence, strong separability implies independence among groups or groupwise independence. The Hessian matrix V in this case is block-additive ($U_{ij} = 0$, $i \; \varepsilon \; G_s$, $j \; \varepsilon \; G_r$, $r \neq s$). Formally, this concept implies additivity among groups and only implies additivity when each group comprises a single commodity. It is clear that weak separability is a less stringent assumption than strong separability. In fact, strong separability or groupwise independence implies weak separability. Under weak separability, a unique intergroup coefficient exists for each pair of groups, but under strong separability, the intergroup coefficient is the same for all groups. Again, the cross-price elasticities among commodity groups are proportional to the relevant income elasticities. With knowledge of all the income elasticities and one intergroup coefficient, all cross-price elasticities between commodity groups follow by direct computation.

Pearce separability. A utility function is Pearce separable with respect to the partition in question, if the marginal rate of substitution between any two commodities belonging to the same subset is independent of the consumption levels of all other commodities, including other commodities within the same subset. Mathematically,

$$\partial(U_i/U_j)/\partial q_k = 0 \quad \text{for all } i, \quad j \; \varepsilon \; G_s, \quad k \neq i, j \quad (1.19)$$

Pearce's assumption requires that where there are more than two commodities in a group, any two commodities within that group must be in "neutral want association" with all other commodities. No utility relation exists between two goods i and k different in any special way from the utility relation between goods j and k. However, since a nonzero relationship can exist, Pearce's concept of neutral want association differs from Frisch's concept of want independence. Neutral want association, which embodies the concepts of weak and strong separability, implies a utility function of the form

$$U(q_1, \ldots, q_n) = F\{U^1[f_1^1(q_1^1) + \ldots + f_{n_1}^1 (q_{n_1}^1)] \, , \ldots ,$$

$$U^S[f_1^S(q_1^S) + \ldots + f_{n_S}^S (q_{n_S}^S)]\}. \quad (1.20)$$

The utility function is weakly separable among groups and strongly separable within groups. For Pearce separability, similar to other concepts of separability, each cross-price effect can be expressed in terms of two income effects and an intergroup coefficient which holds between each pair of groups and is constant over all pairs of goods taken from the two groups.

In summary, separability of commodities within a utility function assumes that the ratio of marginal utilities of a pair of commodities i and j is invariant to the level of consumption of a third commodity k. Various types of separability assume only that the marginal utilities of goods i and j are changed equally because of a change in the consumption of the commodity k. Johnson, et al. (1984, 57-58) illustrate the concepts of weak, strong, and Pearce separability through the use of the direct quadratic utility function.

In practice, it is next to impossible to observe the marginal utilities to identify the groups and the nature of the separability. The identification of homogeneous groups generally rests on two approaches: (1) factor and cluster analysis, and (2) intuition. Both approaches are not unique across researchers because the commodity groupings are often subjective. Additive preferences, almost additive preferences, and separability conditions are intuitively tenable relationships which effectively reduce the dimension of the parameter space while

still preserving most of the simultaneity of the system. Table
1.2 provides a summary of the number of independent parameters
that require estimation under each of the above conditions.

TABLE 1.2. Number of independent parameters that require estimation
under conditions of additive preferences, almost additive preferences, and various types of separability

Condition	No. independent parameters
Classical conditions	$(n^2 + n - 2)/2$
Additive preferences	n
Almost additive preferences	$(n + 1) + (\sum\limits_{i=1}^{n} k_i)/2$
Weak separability	$S(S - 1)/2 + \sum\limits_{G=1}^{S} [n_G(n_G + 1)/2] - 1$
Strong separability	$[\sum\limits_{G=1}^{S} n_G(n_G + 1)]/2$
Pearce separability	$2n$

Note: n = total number of commodities; k_i = number of interactions
for the ith commodity with other commodities, i = 1, ..., n; S = number
of separate commodity groups; n_G = number of commodities in the Gth
group, G = 1, ..., S.

EXTENSIONS OF NEOCLASSICAL THEORY

The neoclassical model of consumer behavior has several
limitations. The purpose of this section is to discuss extensions of the neoclassical theory to take into account socioeconomic, demographic, sociopsychological, and nutritional
factors, choices under risk and uncertainty, plus new commodities, multiperiod consumption decisions, and time.

Socioeconomic, Demographic, Sociopsychological, and Nutritional Factors

To analyze the effects of price and income on quantities
demanded, it is necessary to isolate the effects of other factors which primarily affect taste and preference such as psychological, sociological, cultural, nutritional, and regional factors that determine the level of consumption of a given commodity. Indeed, there exists a vector of determinants other than
price and income that precipitates shifts in the demand vector
of quantities. Such determinants, however, have seldom been
employed in the context of systems of demand equations.

Socioeconomic and demographic characteristics such as family
size and composition, race, age, sex, religion, education, place
of residence, and geographical location have traditionally been
used as arguments in consumer demand studies. Family size and
composition have been standardized by the use of consumer unit
scales (Buse and Salathe 1978; Muellbauer 1974). Recent developments in consumer demand theory emphasize sociopsychological
attributes (Bayton 1977). Different sets of consumers have

different structures of needs, expectations, perceptions, and attitudes. Since sociopsychological characteristics of consumers are likely to influence the demand for commodities, the linkage between consumer motivations and purchase behavior is crucial. Moreover, nutritional factors such as food and nutrition programs (food stamp program, WIC program, dietary guidelines, and recommended dietary allowances), labeling of foods, and food safety regulations may exert a notable influence on consumer demand. Presently, unprecedented interest in the subject of nutrition has been expressed by consumer advocate groups, by health food interests, and by the government.

Economists must be attuned to a number of socioeconomic, demographic, sociopsychological, and nutritional factors. In general, there are two ways to proceed to account for the impact of such characteristics: (1) use methods to analyze subsamples of consumers with identical profiles, and (2) make assumptions which relate the behavior of consumers with different profiles. The former approach allows all of the parameters of the demand system to depend on the socioeconomic, demographic, sociopsychological, and nutritional factors, but this approach does not require a specification of the form of the relationships between the demand parameters and the respective characteristics. Under this procedure, the only data relevant to the analysis of consumers with a particular profile are the observations on consumers with that profile. The latter approach permits practitioners to draw inferences about consumers with one profile from observations on the behavior of consumers with different profiles.

Socioeconomic, demographic, sociopsychological, and nutritional factors can be introduced into any system of demand equations in several ways. One approach, advanced by Lau, et al. (1978), and by Parks and Barten (1973), introduces these characteristics, say α_1, ..., α_p, either continuous or discrete, directly into the direct or indirect utility function. The demand functions which arise then depend in part on the vector of the various factors. In short, these researchers specify a utility maximization model for consumer behavior which takes into account the impact of differential composition of consumers in a general way.

Alternatively, Pollak and Wales (1981) describe four general procedures for incorporating the aforementioned factors into demand systems: (1) translating, (2) scaling, (3) Gorman and reverse Gorman specifications, and (4) the modified Prais-Houthakker procedure. As Pollak and Wales state, "each procedure replaces the original class of demand systems by a related class involving additional parameters and postulates that only these additional parameters depend on the socioeconomic, demographic, sociopsychological, and nutritional variates. The specification is completed by postulating a functional form relating these newly introduced parameters to the respective variables" (p. 1534).

Choices under Risk and Uncertainty

The neoclassical theory of consumer behavior does not include analyses of choice patterns under uncertain situations.
Neoclassical theory assumes the consumer possesses perfect
knowledge. In reality, consumers have imperfect information
with regard to income, commodity prices, commodity availability,
and commodity quality. Decision-making models which account for
risk and uncertainty have been developed by Friedman and Savage
(1948), Von Neumann and Morgenstern (1947), and Luce and Raiffa
(1965).

Von Neumann and Morgenstern showed that if consumer behavior
satisfies certain crucial axioms (the complete-ordering axiom,
the continuity axiom, the independence axiom, the unequal-prob-
ability axiom, and the axiom of complexity), it is possible to
construct a utility index that describes consumer preferences
numerically and predicts choices in uncertain situations. A
complete utility index can be derived by successively confront-
ing the consumer with various choice situations involving certain outcomes on the one hand and probabilistic combinations of
uncertain outcomes on the other hand. It can be shown that the
consumer who conforms to the five axioms maximizes expected
utility. The expected utility calculation determines the consumer's choices in situations involving risk. If the consumer
faces a set of uncertain prospects, the consumer chooses the
prospect with the highest expected utility.

New Commodities

Consumer reactions to new commodities and to quality variations are notable aspects of consumer behavior. The neoclassical
theory, however, inefficiently accommodates such reactions. In
essence, to take into account new commodities and quality variations, the neoclassical approach requires an expansion of the
commodity space to include changes in the number of commodities
available to the consumer. This expansion explicitly entails
the replacement of the old utility functions with a new utility
function which contains the new and/or modified set of commodities. The optimum choice of commodities proceeds via the maximi-
zations of the new utility functions. If the consumer contin-
ually faces new or modified commodities, the neoclassical model
forces the consumer to incessantly search for optimal choice
sets.

To overcome this limitation and others, Lancaster (1966) has
developed a new approach to consumer theory. The novelty lies
in breaking away altogether from the neoclassical approach to
consumer behavior. The assumptions of the Lancastern framework
are threefold: (1) commodities per se do not give utility to
the consumer; such commodities possess characteristics or properties, and these characteristics generate utility; (2) in
general, commodities possess more than a single characteristic,
and characteristics may be shared by more than one commodity;
and (3) commodities in combination may possess characteristics

different from those pertaining to the commodities separately. The set of characteristics is the same for all consumers. Utility or preference orderings rank collections of characteristics and only rank collections of commodities indirectly through the characteristics that they possess.

Consumption activities consist of a transformation of goods to characteristics. The relationship between the collection of commodities and the consumption activities is $q = Ac$, where q is a $(n \times 1)$ quantity vector of commodities, c is a $(m \times 1)$ vector of consumption activities, and A is a $(n \times m)$ matrix of technical transformation coefficients that relate the type and quantity of goods associated with each consumption activity. Similarly, the relationship between the collection of characteristics available to the consumer and the consumption activities is $z = Bc$, where z is a $(s \times 1)$ vector of characteristics and B is a $(s \times m)$ matrix of technical transformation coefficients that relate each of the consumption activities with characteristics.

The consumer optimization problem under the Lancastern approach is simply

$$\text{maximize } U(z)$$
$$\text{subject to } z = Bc$$
$$q = Ac$$
$$\sum_{i=1}^{n} P_i q_i \le y \qquad (q,\, c,\, z \ge 0) \qquad (1.21)$$

In brief, the model defines the utility function over the set of characteristics, the budget constraint over the set of commodities, and the relationship between commodities and characteristics by A and B.

Within this framework, which constitutes a generalization of the neoclassical theory, dimensional changes in the set of commodities simply alter the number of columns of A. The number of characteristics and the utility function remain the same. In short, the new approach to consumer theory accommodates new commodities and/or quality variations without the requirement of the consumer to remaximize utility. However, despite the theoretical potential of the Lancastern approach to demand theory, the approach necessitates further refinement for empirical applications due to the unobservability of the characteristics z and the nonlinearity of the program in the solution to the consumer optimization problem.

Multiperiod Consumption Decisions

The consumer choice problem for the neoclassical model depends on decisions for a single period of time. However, due to a host of factors such as expectations about future prices and income, the influence of past incomes, and inertia, consumer adjustments to changes in prices and income may not be complete

during the single time period. Hence, an extension of the
static theoretical framework of consumer choices to a dynamic
theoretical framework is in order. The purpose of this section
is to introduce a multiperiod utility index and budget constraint to cover optimization for multiple period horizons.

The multiperiod utility index depends on the vector of
quantities of the commodities over the entire planning horizon.
With n commodities and T time periods, the utility function may
be written as

$$U = U(q_{11}, \ldots, q_{n1}, q_{12}, \ldots, q_{n2}, \ldots, q_{1T}, \ldots, q_{nT})$$
$$(1.22)$$

where q_{it} (i = 1, ..., n, t = 1, ..., T) represents the quantity
of the ith commodity in time period t. The consumer desires to
maximize the level of the index U subject to a lifetime budget
constraint which requires the equality of the present values of
the income and consumption streams.[4] The budget constraint may
be written as

$$\sum_{i=1}^{T} y_t (1 + r_{1t})^{-1} = \sum_{t=1}^{T} \sum_{j=1}^{n} p_{jt} q_{jt} (1 + r_{1t})^{-1} \qquad (1.23)$$

where y_t is the income in period t, p_{jt} is the price of the jth
commodity in period t, and r_{1t} is the marginal market rate of
return for the investment periods 1, ..., T; r_{1t}, in turn, depends on the market rate of interest for the periods 1, ...,
t − 1 (i_1, ..., i_{t-1}).

With the multiperiod utility function and the lifetime
budget constraint, the consumer optimization problem has the
following mathematical representation (Henderson and Quandt
1958):

maximize $L = U(q_{11}, \ldots, q_{nT})$

$$+ \lambda \sum_{t=1}^{T} (y_t - \sum_{j=1}^{n} p_{jt} q_{jt})(1 + r_{1t})^{-1} \qquad (1.24)$$

The first-order conditions of utility maximization are similar
to those for the single-period analysis with some modifications. Commodities are now distinguished by time period as well
as by kind, and discounted prices replace simple prices. The
second-order conditions are the same for the general one-period
analysis. Solving the set of first-order conditions leads to
the set of demand functions

$$q_{it} = q_{it}(p_{11}, \ldots, p_{nT}, i_1, \ldots, i_{T-1}) \qquad (1.25)$$

for i = 1, ..., n; t = 1, ..., T. The consumer demand for the
ith commodity in the tth time period depends on the price of
each commodity in each time period and the interest rates in
periods 1, ..., T - 1.

Time

The neoclassical model essentially ignores the cost of time
in the consumer choice problem. To circumvent this limitation
of the neoclassical theory of consumer behavior, Becker (1965)
provides a theoretical analysis of choice that includes the cost
of time on the same footing as the cost of commodities. The
Beckerian framework provides an integration of production and
consumption into economic analysis.

In this framework, the consumer combines time and market
goods to produce basic commodities. Such basic commodities
directly enter the utility function of the consumer. In short,
the consumer is both a producing unit and a utility maximizer.
Let Z_i (i = 1, ..., n) denote the basic commodities. Denote the
"production functions" with inputs time and market goods to
produce Z_i by

$$Z_i = f_i(q_i, T_i) \qquad (i = 1, ..., n) \qquad (1.26)$$

where q_i is the ith market good and T_i is the ith time input.
The consumer in familiar fashion chooses the best combination of
the basic commodities via the maximization of the utility function

$$U = U(Z_1, ..., Z_n)$$

$$= U(f_1, ..., f_n) = U(q_1, ..., q_n; T_1, ..., T_n) \qquad (1.27)$$

subject to the budget constraint

$$\sum_{i=1}^{n} p_i q_i + L(Z_1, ..., Z_n) = S \qquad (1.28)$$

where S represents the money income achieved if all the time were
devoted to work, p_i corresponds to the price of the ith market
good q_i, and L denotes the total earnings foregone by the
interest in leisure or nonwork activities. Note that L is a
function of $Z_1, ..., Z_n$ since how much is foregone depends on the
set of basic commodities. The money income is "spent" on the
commodities Z_i either directly through expenditures on market
goods or indirectly through the foregoing of income by using
time at nonwork activities rather than at work.

In the Beckerian model, the consumer maximizes utility subject to resource and production constraints in order to determine an optimal bundle of commodities. All market goods are inputs, and hence, the consumer demands for these goods are derived demands analogous to derived demands of the firm for factors of production. With this formulation, the set of demand equations for the market goods depends on money income, the shadow price of time, the factor environment, and the n market good prices.

KEY ISSUES IN DEMAND ANALYSIS

Several important issues in demand analysis generally receive little attention. The purpose of this section is to discuss three of these issues: (1) the choice between complete and partial demand systems, (2) aggregation issues, and (3) the choice between single or multiple market level demand.

Complete and Partial Systems of Demand Relationships

The complete systems approach, empirically, is in an embryonic stage of development. From previous work, the complete systems approach deals with the set of demand functions which describes the allocation of total expenditure among an exhaustive set of consumption categories derived from a "well-behaved" preference ordering. The sum of the expenditures on the consumption categories must equal the total expenditure under consideration. Total expenditure is equal to total income if the exhaustive set of categories of commodities includes a savings (or dissavings) category. This approach provides a conceptual framework to deal with the interdependence of demand for various commodities. This integrationist approach provides information on the degree and nature of the interrelatedness of the demand functions and makes assumptions regarding the interaction of commodities and the nature of utility functions.

In contrast, the partial systems approach may fail to recognize explicitly the interrelationships among commodities. The partial systems approach embodies either analyses of single commodities or subsets of commodities where estimation methodology or analyses of the vertical linkages in the marketing channel from the farm sector to the retail sector may be the prime concern.

Demand analyses for a single commodity or a subset of commodities occur for a number of reasons. Researchers must select analyses on the basis of trade-offs by evaluating among complexity of the model, the theoretical considerations, and the data and time constraints.

First, it is possible that the subset of demand relationships in question forms a reasonably independent set. In addition, the equations for this subset of commodities may assume different functional forms, each of which cannot necessarily be derived from the same utility function. No theoretical basis exists for choices among mathematical functional forms for the

demand relationships. Hence, adherence to the classical theoretical restrictions is not guaranteed.

Second, the single commodity or sector model approach requires the estimation of a small number of parameters relative
to the complete demand systems approach. Even though various
restrictions may be used to reduce the number of parameters to
be estimated, the estimation procedure for complete demand systems almost always requires a rather substantial sample size.
The necessity of a "large" sample size is most restrictive in
terms of the use of time-series data. A long enough time-series
may not be available to estimate the parameters of a complete
demand system with sufficient disaggregation of the consumption
categories. Or, if the time-series data are available, substantial structural changes may have occurred. The difficulties
of specifying or maintaining the classical restrictions of demand theory may be especially substantial if the parameters of
each of the equations are found to vary over time and to change
in different ways.

Third, the estimation task, the cost of estimation, and the
statistical problems of multicollinearity and nonspherical error
structures for complete demand systems are typically more bur-
densome than for partial demand systems. In short, a theoretically attractive specification may not be empirically superior
and may turn out to be unmanageable. If a compromise of the
theory yields an amenable and/or empirically almost equivalent
system, the choice for the partial systems approach is clear for
those who want to apply economic reasoning to explain consumer
behavior.

The major contributions and usefulness of complete demand
systems are twofold: (1) the generation of a massive volume of
empirical results consistent with the economic theory of consumer behavior, and (2) information to test hypotheses about
restrictions directly obtainable from the economic theory of
consumer behavior.

Complete demand systems usually generate estimates of
own-price elasticities (compensated or uncompensated), cross-
price elasticities (compensated or uncompensated), income
elasticities, and marginal budget shares of all commodities in
the set. The additive complete demand systems provide estimates
of welfare indicators such as the income elasticity of the
marginal utility of income (w), the income flexibility (ϕ),
and the marginal utility of income (λ). A particular demand
system permits the estimation of the marginal propensity to
consume and in this sense links consumer demand theory with
macroeconomic theory. Some system specifications also provide
estimates of elasticities of substitution among goods and of
subsistence levels of consumption or expenditure for individual
goods. The use of complete demand systems has provided information for testing postulates theoretically derived from the
economic theory of consumer behavior under the condition that
the consumer maximizes satisfaction subject to a budget con-

straint.   Further, the use of complete demand systems has
permitted the testing of hypotheses about various groupings of
goods according to alternative separability rules. Thus, the
complete systems approach has provided economists with an
experience of testing and applying a sophisticated set of
theoretically derived restrictions to actual data, an opportunity not often available to social scientists (Brown and Deaton
1972).

In general, there are two ways to formulate complete demand
systems: (1) specify a particular direct or indirect utility
function, and (2) specify the functional form of the demand
equations directly and impose the classical and modern theoretical restrictions. The neoclassical theory is not much help in
specifying the functional form of a system of demand relationships. However, all the equations in the system have the same
functional form since constraints across equations need to be
imposed. The solution of the constrained utility maximization
problem generates a set of demand relations whose functional
form depends on the form of the utility function. For some
utility functions, the explicit functional form of the demand
functions has been derived. In these instances, all classical
restrictions are automatically satisfied globally, and further,
the distinctive utility function chosen usually yields additional restrictions.

This procedure is not without problems, however. The number
of known and well-behaved utility functions is very limited, and
the derivation of demand equations is not always possible.
There may be doubts about the suitability of adopting a particular specification, and in many cases, the derived demand
relations turn out to be highly nonlinear in their parameters.
Hence, in some circumstances, economists have preferred to work
with an arbitrary but manageable functional form enforced on the
behavioral relations to be estimated, imposing constraints which
insure their theoretical plausibility. However, the theoretical
restrictions are only enforced at some local set of coordinates,
often the sample means.

Several versions of empirical static demand systems exist:
(1) the translog systems (Christensen, et al. 1975), (2) the
Rotterdam system (Barten 1969, Theil 1975), (3) the addilog
systems (Houthakker 1960), (4) the constant elasticity of demand
system (CEDS) (Capps, 1979), (5) the linear expenditure system
and various generalizations (Stone 1953; Brown and Heien 1972;
Lluch and Williams 1975; Blackorby, et al. 1978; Pollak and
Wales 1978), (6) the Australian models (Leser 1961, Powell
1974), (7) the multinomial logit model (Tyrrell and Mount 1978),
and (8) the almost ideal demand system (AIDS) (Deaton and
Muellbauer 1980a).   The Rotterdam system, the CEDS, the
Australian models, the multinomial logit model, and the AIDS are
directly specified demand systems, while the remaining systems
are based on utility function approaches. A description of the
various strengths and weaknesses of the functional forms is

beyond the scope of this chapter. (See, for example, Johnson, et al. 1984). However, discussions of several of the demand systems will follow in later chapters.

Neoclassical theory provides a static interpretation of consumer behavior. However, dynamic specifications of complete systems and partial systems are necessary to provide a scope for adjustment, habit, and inventory features of consumer behavior. The state adjustment model introduces dynamics into complete demand systems wherein quantities purchased depend on either physical stocks of goods or psychological stocks of habits (Houthakker and Taylor 1970). A second way of capturing dynamic features of consumer behavior in the structure of demand systems is to use a dynamic utility function which takes into account changes in tastes and preferences (Phlips 1972). A third way of incorporating dynamics into complete demand systems is to cast the problem into a control theory framework in which the consumer attempts to maximize a discounted utility function subject to wealth and stocks constraints (Lluch and Williams 1975). A fourth way of introducing dynamics into complete demand systems is to define particular parameters to be functions of consumption in previous periods (Pollak and Wales 1969; Brown and Heien 1972). For partial demand systems, distributed lag models and recursive systems have been especially useful to incorporate the dynamic elements. For either partial or complete demand systems, the introduction of adjustment, habit, and inventory features of consumer behavior serves to distinguish between short-run and long-run price and income elasticities and the path of adjustment to other variables.

Aggregation Issues

Neoclassical theory pertains to a single individual, and so the economic theory of consumer behavior is micro in nature. However, practical application of the theory involves econometric estimation and hypothesis testing based on available data. Either the data are lacking for sufficiently elementary commodities, time units, and individual consumers, or practitioners use up too many degrees of freedom when taking into account all detailed data explicitly. Hence, aggregation is typically unavoidable in empirical work on complete and partial demand systems, and analyses are subject to aggregation problems. Aggregation theory deals with the transformation of micro relationships to macro relationships (Thiel 1954; Green 1964).

Basically, there are three types of aggregation problems: (1) aggregation over individuals, (2) aggregation over commodities, and (3) aggregation over time. According to Green "aggregation is a process whereby a part of the information available for the solution of a problem is sacrificed for the purpose of making the problem more easily manageable" (p. 3). Usually in empirical work, the behavior of a single individual is not interesting; economists generally study the behavior of the market, the aggregation of all individual consumers. Thus, "real world" consumer demand generally consists of the aggregate demand for a number of consumers.

The rationale for reliance of the theory of demand for an individual consumer to provide restrictions on a set of demand functions at the aggregate level may be weak. In general, there exists no collective preference ordering, and hence, no representation in the form of an aggregate utility function. Further, the aggregation over consumers may result in the untenability of the assumption of the exogeneity of commodity prices. Therefore, supply relationships may need to be considered explicitly in order to handle the simultaneity between consumer prices and consumer purchases. Nevertheless, Barten (1974) argues that this type of aggregation process introduces errors of only negligible order of magnitude, and presents a case in favor of the analogy between the properties of individual demand equations and aggregate demand equations.

Often, it is crucial to aggregate individual commodities to commodity groups and to aggregate over time periods for various reasons. However, difficulties of grouping homogeneous goods, obtaining price and quantity measures of aggregates are common problems. In addition, there is the difficulty of interpreting the estimated coefficients of aggregate goods. However, Houthakker and Taylor concluded that of all errors likely to be made in demand analysis, the aggregation errors are the least troublesome.

Multiple Market Level Demand

The consumer makes purchase decisions on the basis of prices, income, and other factors for commodities at the retail level. However, the commodities typically funnel through a number of channels before reaching the consumer. In short, a number of intermediaries may exist between the producer and the consumer. Emphasis on this vertical market structure has long been popular for agricultural economists. The importance of this area of demand research stems from the ability to analyze the impact of factors in a particular level of the marketing channel on participants in other levels. To illustrate, decisions at the producer level may have impact on the consumer at the retail level, and similarly, consumer policies and programs may have impact on the producer at the farm level. In addition, behavior at various intermediary points in the marketing channel may influence behavior at both the retail level and the farm level.

Gardner (1975) and Heien (1980) set forth the basic determinants of the linkage of the retail and farm levels in the vertical market structure in a framework which consists of six behavioral relationships: (1) the consumer or retail demand relationship, (2) the retail supply relationship, (3) the intermediary demand relationship, (4) the intermediary supply relationship, (5) the demand relationship at the farm level, and (6) the supply relationship at the farm level. There may be situations where the consumer demand portion of the model may be treated separately from the rest of the vertical market structure. For some commodities, the model determines the retail price and quantity, the farm price and quantity, and the intermediary price and quantity (marketing services). For time periods where inventory change is small in relation to total demand, to assume equality of supply and demand relationships for locating the respective equilibrium price and quantity combinations for each level of the market structure is realistic. However, as the time period under consideration becomes shorter and shorter, disequilibrium occurs since time is required for each of the marketing levels to clear.

The analysis by Gardner has little to say concerning the time path from one equilibrium point to another. The analysis by Heien not only puts forth a theory consistent with the static framework of Gardner but also attempts to describe the dynamics of the vertical market structure. To develop an empirical econometric model of this structure, researchers typically employ either a simultaneous systems approach, a recursive systems approach, or variations thereof. However, both Gardner and Heien develop the theoretical models in terms of a single commodity. The theoretical model of the consumer demand portion of the vertical market structure recognizes the interrelationships of all commodities. The obvious call is for the theoretical treatment of several commodities within the vertical system.

SUMMARY

This chapter presents the theoretical framework for empirical work on consumer demand. Emphasis is on concepts of neoclassical theory, extensions of neoclassical theory, and key issues in demand analysis. This chapter provides much of the framework from which the demand analyses described in later chapters proceed. To illustrate, Chapters 4 through 8 deal with various applications and types of complete demand systems. Chapter 9 entails a discussion of partial systems of demand analyses. Chapter 10 centers on socioeconomic, demographic, and psychological variables in demand analyses, while Chapters 11 through 13 focus on nutritional factors in demand analyses.

First however, in Chapters 2 and 3, attention is devoted to the input data available for various demand models. In addition to limitations of specific models, there are also significant limitations regarding data that the analyst should recognize.

NOTES

1. The Debreu theorem does not, however, imply uniqueness of the utility function. Any monotonic transformation of the utility function may represent equally well the same preference ordering. Such transformations bring to light the notions of

cardinal measures and ordinal measures. Cardinal measures are
invariant only under linear transformations, while ordinal
measures are invariant under linear and nonlinear transformations.

2. The Slutsky symmetry conditions provide n(n - 1)/2 restrictions, the homogeneity and Cournot aggregation conditions
each provide n restrictions, and the Engel aggregation condition
provides a single restriction. However, since it is possible to
derive the homogeneity restrictions or the Cournot aggregation
restrictions from the other classical restrictions, the number
of independent restrictions is $(n^2 + n + 2)/2$.

3. The basic idea is to partition the elements of the
commodity bundle into different groups. Under this scheme,
budget allocation proceeds in a stepwise fashion such that the
consumer first allocates total expenditure (or income) among the
groups, and then the consumer allocates these optimal group
expenditures among the individual commodities within the respective group.

4. For particular periods, the income streams may not coincide with the consumption streams. However, through borrowing
and lending the consumer may reconcile the two streams.

REFERENCES

Allen, R. G. D., and A. L. Bowley. 1935. Family expenditure.
    London: Staples Press.
Allen, R. G. D., and J. R. Hicks. 1934. A reconsideration of
    the theory of value. Economica 1:196-219.
Barten, A. P. 1964. Consumer demand functions under conditions
    of almost additive preferences. Econometrica 32:1-38.
_____. 1969. Maximum likelihood estimation of a complete sys-
    tem of demand equations. Eur. Econ. Rev. 1:7-73.
_____. 1974. Complete systems of demand equations: Some
    thoughts about aggregation and functional form. Rech. Econ.
    de Louvain 40:1-18.
_____. 1977. The systems of consumer demand functions ap-
    proach: A review. Econometrica 45:23-51.
Bayton, J. A. 1977. Needed research of the impact of socio-
    psychological factors on food demand. In Food demand and
    consumption behavior, ed. Robert Raunikar. S-119 Southern
    Regional Research Committee and the Farm Foundation. Athens:
    Univ. of Georgia.
Becker, G. S. 1965. A theory of the allocation of time. The
    Econ. J. 75:493-517.
Bieri, J., and A. de Janvry. 1972. Empirical analysis of demand
    under consumer budgeting. Giannini Found. Monogr. No. 30.
    Univ. of California, Berkeley.
Blackorby, C., R. Boyce, and R. R. Russell. 1978. Estimation of
    demand systems generated by the Gorman Polar Form: A
    generalization of the S-branch utility tree. Econometrica
    46:345-65.

Brandow, G. E.   1961.   Interrelations among demands for farm
    products and implications for control of market supply. Pa.
    Agric. Exp. Stn. Bull. No. 680.
Brown, J. A. C., and A. S. Deaton. 1972. Surveys in applied
    economics: Models of consumer behavior. Econ. J. 82:
    1145-236.
Brown, M., and D. Heien. 1972. The S-branch utility tree: A
    generalization of the linear expenditure system. Economet-
    rica 40:737-47.
Buse, R. C., and L. E. Salathe. 1978. Adult equivalent scales:
    An alternative approach. Am. J. Agric. Econ. 60: 460-68.
Capps, O., Jr. 1979. The impacts of selected nonfoods, foods,
    socioeconomic and demographic characteristics on the
    decision to purchase various meats and seafoods for home
    consumption. Ph.D. diss., Virginia Polytechnic Inst. and
    State Univ.
Christensen, L. R., D. W. Jorgenson, and L. J. Lau. 1975.
    Transcendental logarithmic utility functions. Am. Econ.
    Rev. 65:367-83.
Deaton, A., and J. Muellbauer. 1980a. An almost ideal demand
    system. Am. Econ. Rev. 70:312-26.
_____. 1980b. Economics and consumer behavior. Cambridge:
    Cambridge Univ. Press.
Debreu, G. 1959. Theory of value: An axiomatic analysis of
    economic equilibrium. Cowles Monogr. No. 17. New York:
    Wiley.
Friedman, M., and L. J. Savage. 1948. The utility analysis of
    choices involving risk. J. Polit. Econ. 56:279-304.
Frisch, R. 1959. A complete scheme for computing all direct
    and cross-demand elasticities in a model with many sectors.
    Econometrica 27:177-96.
Gardner, B. L. 1975. The farm-retail price spread. Am. J.
    Agric. Econ. 57:399-409.
George, P. S., and G. A. King. 1971. Consumer demand for food
    commodities in the U.S. with projections for 1980. Giannini
    Found. Monogr. No. 26. Univ. of California, Berkeley.
Goldberger, A. S. 1967. Functional forms of utility: A review
    of consumer demand theory. Univ. of Wisconsin, Systems
    Found., Methodology and Policy Workshop, Pap. 6703.
Goldman, S. M., and H. Uzawa. 1964. A note on separability in
    demand analysis. Econometrica 32:387-98.
Green, H. A. J. 1964. An aggregation in economic analysis: An
    introductory survey. Princeton, NJ: Princeton Univ. Press.
Heien, D. M. 1980. Markup pricing in a dynamic model of the
    food industry. Am. J. Agric. Econ. 62:10-18.
Henderson, J. H., and R. E. Quandt. 1958. Microeconomic theory:
    A mathematical approach. New York: McGraw-Hill.
Hicks, J. R. 1936. Value and capital. Oxford: Oxford Univ.
    Press.
Houthakker, H. S. 1960. Additive preferences. Econometrica 28:
    244-57.
_____, and L. D. Taylor. 1970. Consumer demand in the United

States 1929-1970. 2nd ed. Cambridge, MA: Harvard Univ. Press.

Johnson, S. R., Z. A. Hassan, and R. D. Green. 1984. Demand systems estimation methods and applications. Ames: Iowa State Univ. Press.

Lancaster, K. 1966. A new approach to consumer theory. J. Polit. Econ. 74:132-57.

Lau, L. J., W. L. Lin, and P. A. Yotopoulos. 1978. The linear logarithmic expenditure system: An application to consumption-leisure choice. Econometrica 46:843-68.

Leser, C. E. V. 1961. Commodity group expenditure functions for the United Kingdom, 1948-57. Econometrica 29:24-32.

Lluch, C., and R. Williams. 1975. Consumer demand systems and aggregate consumption in the U.S.A.: An application of the extended linear expenditure system. Can. J. Econ. 8:49-66.

Luce, R. D., and H. Raiffa. 1965. Games and decisions: Introduction and critical survey. New York: Wiley.

Muellbauer, J. 1974. Household composition, Engel curves, and welfare comparisons between households: A quality approach. Eur. Econ. Rev. 5:103-22.

Parks, R. W., and A. P. Barten. 1973. A cross-country comparison of the effects of prices, income, and population composition on consumption patterns. Econ. J. 83:834-52.

Pearce, I. F. 1964. A contribution to demand analysis. Oxford: Oxford Univ. Press.

Phlips, L. 1972. A dynamic version of the linear expenditure model. Rev. Econ. and Stat. 54:450-58.

_____. 1974. Applied consumption analysis. Amsterdam: North-Holland.

Pollak, R. A., and T. J. Wales. 1969. Estimation of the linear expenditure system. Econometrica 37:611-28.

_____. 1978. Estimation of complete demand systems from household budget data: The linear and quadratic expenditure systems. Am. Econ. Rev. 68:348-59.

_____. 1981. Demographic variables in demand analysis. Econometrica 49:1533-51.

Powell, A. A. 1974. Empirical analytics of demand systems. Lexington, MA: Heath.

Samuelson, P. A. 1965. Economic forecasting and since. Mich. Q. Rev. 4:274-80.

_____. 1966. Foundations of economic analysis. Cambridge, MA: Harvard Univ. Press.

Schultz, H. 1938. The theory and measurement of demand. Chicago: Univ. of Chicago Press.

Slutsky, E. E. [1915] 1952. On the theory of the budget of the consumer. Trans. Olga Ragusa. In Readings in price theory, ed. G. J. Stigler and K. E. Boulding. Homewood, IL: Richard D. Irwin, Inc.

Stone, J. R. N. 1953. The measurement of consumer's expenditure and behavior in the United Kingdom. Cambridge: Cambridge Univ. Press.

Strotz, R. H. 1957. The empirical implications of a utility

     tree.   <u>Econometrica</u> 25:269-80.
Theil, H. 1954.   <u>Linear</u> <u>aggregation</u> <u>of</u> <u>economic</u> <u>relations</u>.
     Amsterdam: North-Holland.
_____. 1975.   <u>Theory</u> <u>and</u> <u>measurement</u> <u>of</u> <u>consumer</u> <u>demand</u>, <u>Volume</u>
     <u>I</u>. Amsterdam: North-Holland.
Tyrrell, T., and T. Mount.   1978.   <u>An</u> <u>application</u> <u>of</u> <u>the</u> <u>multi-</u>
     <u>nomial</u> <u>logit</u> <u>model</u> <u>to</u> <u>the</u> <u>allocation</u> <u>of</u> <u>budgets</u> <u>in</u> <u>the</u> <u>U.S.</u>
     <u>consumer</u> <u>expenditure</u> <u>survey</u> <u>for</u> <u>1972</u>.   Cornell Univ. Agri.
     Exp. Stn. Bull. No. 20.
Von Neumann, J., and O. Morgenstern.   1947.   <u>Theory</u> <u>of</u> <u>games</u> <u>and</u>
     <u>economic</u> <u>behavior</u>. Princeton, NJ: Princeton Univ. Press.
Wold, H. and L. Jureen.   1953.   <u>Demand</u> <u>analysis</u>:   <u>A</u> <u>study</u> <u>in</u>
     <u>econometrics</u>. New York: Wiley.

CHAPTER 2

Data Sources
for Demand Analyses

INTRODUCTION

A wide variety of data series have been designed, developed, and maintained by public and/or private sources in the United States. Most, however, were not designed with demand analysis use in mind. Therefore, it is helpful to know more about the characteristics and quality of the available data series. Data systems are comprised of three dimensions. One is the time frame, that is, the periodicity and continuity of the data. A second is the complexity of information: is it highly aggregative or does it provide detailed socioeconomic, cross-sectional, market segment data? Is information provided by product subcategories within the major product classes? The third dimension pertains to the marketing level. Do the data represent farm, wholesale, or retail level transactions? These classifications require further exploration. In this chapter, the implications of public versus private data systems, the matter of data accuracy and the nature of different kinds of data will be considered first. The chapter concludes with a detailed discussion on various data sources and systems that are currently available for demand analyses.[1]

Public versus Private Data Systems

The conceptual purpose and design from which public and private data systems evolve are fundamentally different. Public systems were intended to provide information useful to the making of government policy decisions. Consequently, these data are prone to measure quantity flows and related prices for very broad industry and/or major subsector categories of products. Even though the data come from reports of a sample of individual firms, the sample size permits only national level aggregate totals. The so-called public data, whether developed by governmental agencies or universities, is in the area of public domain. Consequently, with some exceptions involving confidentiality,

they are available to everyone. Data for five or fewer firms in
any given category are generally withheld to prevent possible
inferences about business volume of individual companies.

Private data systems, in contrast, are created specifically
to provide individual firm or brand data and are designed to
assist individual firm management decisions. These systems are
designed and operated, as a business, by independent data
development firms. Costs of obtaining and summarizing the data
are paid for through sales of the data to private business
clients and, in some cases, to government agencies. If sales
revenue fails to support the total data system it all col-
lapses. The same is generally true for any individual product
or group therein. To university researchers, even summary data
by product category are seldom provided except on a fee basis,
and fees usually are much too high for university project
budgets.

Quality of the data, public versus private, deserves a
researcher's attention. Public data that are collected and
reported by government agencies, either continuous or periodic,
are usually collected under admirable survey designs. Product
or service are clearly specified as to items covered. Also,
sampling methods satisfy professional standards. However,
one-time survey data for specialized studies, even from government sources, should be closely scrutinized.

Private data vary in quality. Several national private data
services equal the best in any domain. But cost-saving exped-
iencies, or lack of professionalism, sometimes affect private
data quality. Therefore, reviewing methodology for a specific
data series or one-time set is time well spent. Unfortunately,
the omission of research methodology in data or research reports
is a common occurrence, and for data quality evaluation special
methodology reports must be requested.

Though neither in the public nor private domain, a number of
associations maintain their own relevant data banks which may be
available upon special request. Research foundations with data
systems fall into this category. Individual business firms
generate their own data concerning sales and prices as part of
their on-going marketing information systems, but access to
these is limited because of their proprietary nature. At times
however, data may be obtained if it is to be combined with data
from other firms and encoded to protect against revealing
individual firm behavior. Thus, researchers are confronted with
a wide, but at the same time, restricted market coverage spectrum of data having great variance in its quality and accessibility. Yet, compared to the situation in most other nations,
the U.S. data bank has no equal.

Data Accuracy

The accuracy of the data, be it related to market quantities
or prices, depends on the completeness of information
obtained. Unless the total universe of price and quantity

information is included, which is a rare phenomenon, sampling is essential. If sampling criteria are fulfilled, statistical methods provide the basis for measurement of accuracy.

Two caveats apply, however. Those familiar with sampling know that though probability samples are designed, seldom is one achieved. Noncooperation from some surveyed members may lead to inherent data biases. For example, large farms are more likely to consistently supply data to the USDA (because they are often better organized to do so) than are smaller farms. This may result in a sample which is overrepresented by larger and more efficient enterprises.

The other caveat is that in consumer surveys, a diary is usually used as the survey instrument for data collection. Recall or recording errors are often encountered. Few studies concerning data accuracy with respect to different survey methods have been published. An early published study by Levine and Miller (1957) sheds some light on this matter. Their study suggests that results may vary from 10 percent to 25 percent among alternative data collection methods. So long as inherent biases are consistent, the problem from the demand analyst's viewpoint is minimized. But the errors may be additive rather than compensating, particularly among products of widely varying consumer purchase cycles or incidence.

No consensus exists as to how accurate data should be for inclusion in demand analyses. However, an error of 5 percent, or less, would be desirable but probably seldom achieved. Neither can it be fully proved and ascertained if such a level of accuracy is attained because of the complexities associated with the measurement of economic phenomena. Furthermore, in most instances data on a disaggregated basis are not available, except where special consumer panels or surveys are conducted. A considerable chasm occurs between the data needed for an optimum theoretical demand model and that available for application. As shall be seen in the discussion of disappearance data, quantity estimates are often total estimates based on subestimates, and errors in the subseries may be compounding rather than offsetting.

Several key issues are associated with creating and maintaining data series at the consumer level. First is the problem of determining which product(s) within a product family should be selected to represent an entire family of items. The assumption that the item selected represents the subset of items on a continuing basis should be validated. The longer the time frame, the greater the likelihood that ratios within the subset may change, and leverage always exists to reduce the number of representative items selected due to cost consideration of operating the data system. Therefore, the data that is available may not completely satisfy the objectives of demand analysis systems, and measurement difficulties emerge.

A second issue relates to physical changes in a priced product over time. New processing technology leads to product improvement so that the same item represents greater utility but is priced as though it is the same item. Such modifications may permit a lower price or require a higher price. If the index philosophy is to price the most commonly purchased items and ignore product modifications, slippage between the price-utility ratio may occur. However, the establishment of price linkage between the old item and the new item is an approach which may be used in a particular data series in order to circumvent some of the slippage.

The foregoing discussion provides some indication of the conceptual data versus available data differences encountered in demand analyses. Therefore, data series should not be accepted without careful investigation of their definition and the methodology guiding their procurement and calculation. More specific attention will now be given to discussion of two major data sets which can exemplify the nature and extent of the foregoing problems. Considered first is the widely used disappearance data system. Second, the consumer expenditure surveys are evaluated.

Data Time Frames

Most econometric models benefit from an input of continuous market coverage data. Is that input available?

Most of the data available for demand analyses may be classified into one of three major categories with respect to time frames in which the data were collected: (1) continuous survey, (2) periodic survey, and (3) one-time special market survey. Data from a continuous survey, if strictly defined, cover successive total time periods without gaps, and summary totals are given at prescribed intervals (e.g., monthly milk sales data). A periodic survey, in contrast, gathers and reports data regularly but pertains only to subsets of time. Two examples are the monthly Bureau of Labor Statistics Retail Consumer Prices and USDA's Agricultural Prices which relate only to the fifteenth day of each respective month. Many econometric models employ the periodic data and make an inferential assumption that the data are continuous. A one-time special survey is designed to provide data for solving a specific marketing problem, usually of a short-run nature. For example, surveys of consumer attitude toward a new product development in a few selected test markets generally fall into this category.

Aggregative versus Cross-Sectional or Market Segment Data

Aggregative data are stand-ins for either one or all of the following: multiple products, across markets, or across time periods. A large part of government data are of this aggregative type. Prices seldom pertain to the total market volume and usually pertain only to that part moving through a particular level or major marketing channel such as grain elevators. Sales by farmers, however, are to all types of buyers and vary as to qualities or grades involved. Thus, the data collected at a single part of the marketing channels are "stand-in" figures. Commodity sales volumes, one should note, are often a derived figure and not a reported one. Therefore, quantity and price data are not truly matched.

The objective of cross-sectional data is to resolve some of the above problem. It endeavors to provide price and quantity information for specific subclasses of buyers by individual grades sold and other relevant subsets of the total market. Fruits and vegetables offer a good illustration. Processors, or packers, of the fresh canned and frozen forms are respectively different market demand segments exhibiting different demand behavior patterns which usually possess differing demand elasticities. Any interrelationships are measured by cross elasticities.

The type of cross-sectional data one requires depends upon one's econometric model and its objectives. Data pertaining to separate consumer demographic or socioeconomic groups are obviously useful, for example, to the derivation of meaningful income elasticities, but are hard to get. Such data before 1978 were only periodically collected by government agencies. Therefore, current data was seldom available. Data collected by private research firms were usually current but unavailable except at substantial costs. Any psychographics (i.e., attitudinal variables) which are important to commodity marketing as well as to private brand marketing are almost exclusively developed by private research firms and, therefore, are unavailable.

Given the foregoing circumstances, food demand analysts clearly have been without a data sector they would like to have and truly need. Food demand economists' priorities, by nature or training, are to be interested in resultant demand and price rather than its root causes. Therefore, they have not pushed for causative factor measurements.

Another dimension of cross-sectional data detail within a product category also merits attention. Farm level fluid milk quantity sales do not substitute well for retail fluid milk sales since different classes of milk have separate end-markets characterized by different demand behaviors. Similarly, total per capita beef consumption, as measured by disappearance data, is inadequate to a specific demand analysis concerning either fresh or processed meats. Similar problems exist regarding most food products. In an effort to overcome these difficulties, analysts resort at times to some form of derived data series based on a related reported data set.

Marketing Levels of Data and Data Bases
Data can and may be generated at various market levels. Possibilities include those at the farm, at assembly markets, from commodity dealers at major wholesale central markets, from processors, wholesalers, retailers, and/or consumers. Demand elasticities usually differ among the various marketing system

levels.   The appropriate price and quantity data to a specific
market level need to be matched by the researcher.   One should
also determine whether imports or exports are included or
excluded from the sales or disappearance data.   Exports histor-
ically are highly variable.   Concerns about these aspects of
data coverage lead to the question of definitions for the
respective data series.

Importance of Definitions
     Data definitions cannot be expected to automatically fit
those specified in analytical models.   A myriad of classification or definitional questions arise when data generating
surveys or systems are designed.   Many econometricians, as data
users, are unduly unaware of the presence of these problems.   A
cardinal rule is to seek and read definitions carefully before
using data in demand analyses.
     Most surveys provide adequate data definitions.   Some have
to be sought separately.   Figures on canned milk may or may not
include dry powdered forms.   Cream sales may include nondairy
versions.   Home food use may include or exclude that in lunches
taken to work or school.   Questions can arise as to whether and
where frozen or dehydrated foods are included.   Are food quantities that are used in combination items such as frozen dinners
included or ignored?
     Per person or per household data should be defined; for
example, are armed forces members included or excluded?   Is
income cash only or does it include income in kind?   These are
but a few of the ever present data definitional issues that
confront demand analysts.
     All of the foregoing questions require evaluation irrespec-
tive of the particular data system through which data are
acquired.   With this background in mind, we can now consider
several data generating systems, and can look for and recognize
their respective strengths and weaknesses.

DESCRIPTION OF DATA SOURCES AND SYSTEMS
     Data systems are developed and designed for specific levels
within the product marketing system.   The marketing system level
selected depends mostly on two factors: the purpose for which
the system was designed and a cost versus accuracy decision
pertaining to the data gathering processes.   Four classes of
systems are most commonly employed.
     One system is targeted at the production point, where it
monitors the initial product flow.   Combined with periodic
inventory information, it provides disappearance data for use in
farm-level demand analyses.   A second system focuses on the
factory or processor level.   A third intercepts somewhere beyond
the processor level but within the wholesale market distribution
system.   The latter two are suitable for wholesale market demand
studies.   A fourth system is targeted on consumers.   Consumers'
purchases may be monitored either from household consumer

surveys or retail store sales audits. Each of these four basic
data systems has its own purpose and methodology.

Disappearance Data Systems
     Despite being aggregative, disappearance data are useful for
some types of demand analyses. These data, when used, have to
be matched with price data from other sources. The following
section will discuss disappearance data and then consider price
data commonly used with them.

Disappearance data. Procedures used in developing disappearance,
supply, and utilization data are fully described in the USDA
publication Major Statistical Series of the U.S. Department of
Agriculture. Only a summary is possible here. Basically,
disappearance quantities are estimated as a residual balance
from estimated total commodity production, stock changes,
imports and exports balances, and nonfood and nonmarket supply
dispositions. If the market in question is in equilibrium and
all variables are measured without error, then disappearance in
the observational period is conceptually equivalent to the
quantity demanded at prevailing prices in the domestic market.
These estimated quantities appear in USDA's Agricultural Statistics as well as in the USDA's Food Consumption, Prices, and
Expenditures statistical bulletin series.
     Crop production data are developed via a system maintained
by the USDA Crop and Livestock Reporting Board. It operates
through a network of individual state offices cooperating with
the respective state departments of agriculture. A periodic
survey is made at the state level of a sample of farmers and
ranchers to obtain raw input data. At times, special surveys
are made nationally direct from Washington, D.C. Inventory
stocks reports are obtained from periodic surveys among processors and/or commercial warehouses (usually by the U.S. Department of Commerce) and from USDA reports from farmers since
farmers may hold unsold on-farm stocks. Stocks data are collected only a few times per year, one coinciding with the end of
the marketing year. Retail and wholesale stocks are not measured, except for that part that is in public warehouse storage.
Nonfood consumption measurement is indicated, where important,
such as potatoes or grains fed to livestock or diverted to
alcohol production. Nonmarket consumption is measured such as
military procurement from domestic food supplies and USDA food
donations to school lunch and welfare programs. Deduction of
these nonmarket amounts is advisable since these end uses
usually are nonprice sensitive and camouflage basic demand
relationships. Imports and exports are reported by the U.S.
Department of Commerce and must be used to determine net domestic market supplies. Disappearance data, therefore, are designed to measure U.S. civilian market supplies.

Price data. Appropriate price data must be matched with disappearance data, if meaningful time-series demand analyses are to

be obtained, since for most commodities price data are available at several levels of the marketing system. The appropriate one should be selected. Farm-level prices as of the fifteenth day of each month appear in the USDA report Agricultural Prices, which in fact is largely based upon a survey of commodity dealers regarding prices paid to farmers. State and national averages are provided. It is important to note that prices reported in Agricultural Prices are representative of the average qualities, rather than any specific grades, sold by farmers. Furthermore, prices are at the first level of the sale. Therefore, the price usually is at grain elevators, livestock auctions, fruit and vegetable packing sheds, or wherever the first sale occurs. Therefore, costs of delivery from the farm, plus grading, packing, or other first handler processing and container costs, to varying degrees, are usually included. The latter usually change over time and can introduce distortion to net price relationships. In a few instances, USDA calculates and reports equivalent on-farm prices. These variations and their definitions, however, are not always clearly indicated.

USDA's Agricultural Marketing Service reports prices by grade and variety for commodities sold on designated central wholesale markets in publications such as Cotton Market News. These reports are product and individual city market specific. But the quantity moving to that city market under contracts and private treaty negotiations are not included and, therefore, those sales are not represented. The assumption also made is that the major markets provide an adequate indication of prevailing prices in other nonreported market areas.

Private price reporting services are also important for some products. The National Provisioner's Daily Market and News Service, commonly called the "Yellow Sheet," for wholesale meat is one. The National Provisioner is a Chicago-based, privately operated firm which assembles and publishes, on a subscription fee basis, daily Chicago closing prices for wholesale meat and meat products. Wholesale level prices are collected for many products by the Bureau of Labor Statistics. Formerly known as "wholesale prices" but now labeled "producer prices," these are, in some cases, classified into raw, partially processed, and finished products. Individual product prices and product group price indexes are reported monthly in publications such as the Bureau of Labor Statistics' Survey of Current Business.

Consumer price level information is gathered and used in computing the (Bureau of Labor Statistics' National Consumer Price Index (CPI). Included are products of the kind and quality purchased by urban wage earners and clerical workers. Prices are not from consumers but are sampled from retail stores. Price indexes are constructed both for food at home and away from home. A major limitation of matching price data with disappearance data is that the quantity data are for aggregated products, such as for pork production, whereas the price data

are for disaggregated items such as pork chops or bacon. The
interaction effect of price flexibilities and elasticities may
vary greatly by each disaggregated item so that their representativeness is not necessarily constant.

The warning is clear that individual disappearance and price
data series should be thoroughly researched in order that full
knowledge is obtained as to their components and other characteristics. Adequacies of the statistical sampling procedures
behind the data will be discussed in Chapter 3.

Household Panel Data

Household consumer panels are another major data source for
food demand analyses. Information regarding consumer purchase
quantities, unit prices paid, and the kind of outlets where
bought are given. The major attributes of panel data are (1)
cross-sectional representation of household characteristics, (2)
continuous reporting over time, and (3) timeliness (quarterly or
monthly data). All three are valuable input data characteristics for demand analysis. Several household panels, some
private and some public, have been established over the years on
either a national or local area basis. We shall now look at
several examples.

The Griffin Panel. The Griffin Consumer Research Panel, a local
area and public panel, was established in 1974 and maintained by
the Georgia Agricultural Experiment Station through June 1981
(Raunikar 1976). A somewhat larger panel operated in Atlanta,
Georgia, from 1956 through 1962 (Purcell et al. 1957).

The Griffin panel contained 120 households selected by a
two-stage stratified random sample. Six household classifications based on household size and income were selected with an
equal number of households in each. After panel reporting
stability was achieved, data retention began the first week of
January 1975, and continued with household replacement as needed
for sample maintenance. Daily food purchases were recorded in
diaries that were mailed in each week.

The Puerto Rico Panel. The Puerto Rico Agricultural Experiment
Station conducts a continuous household panel for the purpose of
studying food demand. The development of the Puerto Rico panel
began in 1977 and started its continuous reporting operation the
first week of October 1978.

Participating households were obtained from a stratified
cluster subsample of the Master Sample of Households prepared by
the Commonwealth Planning Board of Puerto Rico. The Master
Sample was previously designed for use in various other socioeconomic studies. Households within 44 of the 78 Puerto Rican
municipalities were represented. During the initial survey year
(March 1977 to February 1978) a special survey was made every
four months. During these three household visits, model forms
were provided as training instruments to assess the homemaker's

ability to keep correct diaries. A final selection of a household panelist was based upon the following criteria: (1) the
homemaker's willingness to continue in the consumer panel, and
(2) the quality of the returned training forms. These organizational details are noted to provide valuable insight into the
sampling problems and inevitable presence of some sample bias.

Homemakers in households that became panel members were
compensated $1 per diary turned in that met specified reporting
requirements. Each year the payment increased by $0.25. At the
conclusion of the panel's first year, an average of 127 households participated each quarter. That generated a total of
5,776 diaries for the year or an average of 45 diaries per
household. About 33 food items were reported per diary per week.

The Puerto Rican version of the U.S. food stamp program
began in Puerto Rico in July 1974. In 1980, nearly 50 percent
of the Puerto Rican households received food stamps. The Puerto
Rico panel data are of special interest to food demand analysts
for that reason.

MRCA Panel. Perhaps the oldest nationwide U.S. consumer household panel is the one operated by the Market Research Corporation of America. A private data service, it was originally
established as Industrial Surveys in 1939. The name was changed
to MRCA in the 1940s. Its nationwide sample of approximately
7,500 households is stratified to provide geographic region and
major city market data. Though originally developed as a quota
sample, the panel was converted in the early 1950s to a probability sample basis. Several special purpose panels are also
maintained. By participating, households earn trading points
toward obtaining gifts from a catalogue.

Each household records purchases in a weekly diary. Noted
is item brand name, package size, price, and kind of store where
purchased. The total panel data are summarized weekly and
monthly (four-week periods). Less frequently published reports,
usually for six-month periods, offer summary data by geographic
region, household size, city size, age of household head,
education, and total household income.

MRCA panel data cover food and nonprescription drug items,
though some categories and items are excluded because of lack of
client demand for them. The detailed brand package size and
frequency of purchase information are sold to manufacturing and
marketing firms. Aggregative, or generic, data may be made
available for university use under certain disclosure restrictions, only for the cost of retrieving the information from the
massive tape files.

National Purchase Diary Panel. National Purchase Diary Research
(NPDR) is another private market research firm operating household panels.[2] It too supplies data on a fee basis to
individual firms that utilize it to monitor specific product
markets. NPDR panels cover food away from home (FAFH), textiles

and apparel, wine, toys, gasoline, package goods, and financial
services. Purchase diaries are kept by a nationally representative sample. Detailed purchase information is provided at
regular intervals, enabling an analyst to trace changes in
purchase behavior of particular households or classes of households over time.

One specialized panel operated by NPDR is known as CREST
(Chain Restaurant Eating-out Share Trend). CREST is the NPDR
household survey on FAFH purchase behavior. This quarterly
survey began in September 1975. Meals eaten out or obtained
from commercial eating establishments are reported during a
specified two-week period. School lunches and meals at non-
public and/or nonprofit establishments are excluded.

A major advantage of CREST is the detailed information
provided on household FAFH purchases. Data from the 1972-73 BLS
Consumer Expenditure Survey (composed of separate diary and
interview surveys) and from the 1977-78 USDA Nationwide Food
Consumption Survey are limited mostly to aggregate FAFH expenditures. In some cases the breakfast, lunch, school lunch,
dinner, and snacks totals are delineated. CREST instead provides detailed information regarding each away-from-home meal
occasion. Noted are the establishment name, type of establishment (fast food, cafeteria, etc.), meal occasion, day of week,
price of meal, amount of tip, means of payment (cash, check, or
credit card), food items eaten by each family member, and the
type of coupon used, if applicable. Reported too are the time
and distance traveled to the eating establishment and the
respondent's location before and after the meal. Socioeconomic
and demographic information on the family unit includes family
size and composition, race, income, education, hours of work for
husband and wife, location of residence (census region, state,
market size, and county of residence), and home ownership status.

Each CREST two-week reporting period sample includes 10,000
households, approximately 30,000 individuals, and about 45,000
meal occasions. To improve representativeness, the sample is
stratified by a 288 cell matrix with each cell defined by a
combination of census regions, family income, family size,
housewife age, and whether or not the household resides in a
Standard Metropolitan Statistical Area (SMSA). Report diaries
are mailed to the same 10,000 households each calendar quarter.
In a given year, approximately 25 percent of the households fail
to report all four quarters. Families that drop out of the
survey are replaced using a stratified quota sampling technique
in order to maintain the sample size and proper demographic
representation.

The limitations of FAFH data for demand analyses need to be
recognized. One problem is population representation. As with
many mail surveys, CREST has lower response and higher rates of
attrition from black households, single member households, and
households in the income extremes. Consequently, reporting
households in these sample segments may not be entirely repre-

sentative of their respective population cohorts. As a corrective aid, NPDR has added a special sample of 2,500 single member households. A further limitation is lack of information regarding other food expenditures, such as food at home and school lunches. Therefore, one is preempted within this sample from making comparisons with total household food expenditures or any components. In addition, within CREST there is neither detailed food purchases quantity information nor knowledge of a working wife's wage rate.

Even though the above limitations exist, the data permit a wealth of detail and offer distinctly new avenues of research. Analyses are possible on the effects upon FAFH eating habits of family size and composition, income, race, geographic region, and labor force status of the wife. Influences of type of establishment and/or meal occasion may be delineated. These richly detailed disaggregate data permit an assessment of the impact of changing household characteristics on both the composition and growth of the food service industry. Detailed information by family member permits an examination of who eats what foods and where.

Researchers from outside NPDR's industry clientele have purchased portions of old survey data for use in demand and market analysis. For example, USDA/Economics, Statistics, and Cooperatives Service (ESCS) economists have used data from the wine, textile and apparel, and FAFH surveys to augment and expand their research and information programs in these areas. (See Folwell and Baritelle 1978; Hagar 1979; Smallwood 1980.) In 1979, a special one-year study was conducted for the USDA by NPDR to examine the percentage of income spent on food by households at different income levels. This study, the ESCS Food Expenditure Study, was based on a subsample of households from the on-going FAFH survey. (See Gallo et al. 1980.) Approximately 20,000 households were asked additional questions regarding their aggregate at-home food expenditures. However, the ESCS survey is not considered a substitute for either the 1977-78 USDA Nationwide Food Consumption Survey or the 1972-73 BLS Consumer Expenditure Survey. Rather, it serves as a supplement to those studies and the initiation of the BLS Continuing Consumer Expenditure Survey which began in late 1979.

The Michigan Panel. The Survey Research Center in the Institute for Social Research of the University of Michigan maintains a panel study of income dynamics which provides a uniquely valuable data source for economic research. This panel is one of the few major longitudinal surveys providing data for analyzing the dynamics of household economic behavior. The primary focus is on changes in the economic status of families.

More than 12 years of panel data are available on the same families and their offshoots. Every spring since spring of 1968, heads of families have been interviewed about attitudes, behavior, and economic status. The original survey sample was a

cross section of 2,930 U.S. households and an additional sample
of 1,872 families previously interviewed in 1967 by the Bureau
of Census for the Office of Economic Opportunity making a total
of 4,802 families. A weighting process can make the sample
representative of the U.S. population. Each survey contacts all
households interviewed the previous year and any newly formed
families that contain an adult member of one of the 1968 panel
households, a procedure which, over time, increases the sample
size. Nonresponding households are dropped. Following the 1979
interview wave, the sample included 6,373 families.

Since the 1970s, the Institute for Social Research has
published a series of analytical data volumes entitled Five
Thousand American Families--Patterns of Economic Progress. As
of 1980, the series comprised eight volumes. Ten-volume docu-
mentation series entitled Panel Study of Income Dynamics is also
available.

The surveys have collected very limited information regarding household expenditures. Of interest though to economists
studying consumer food demand are the household expenditure data
for food at home and food away from home. In addition, for a
period of several years, data were collected concerning the
household's food stamp participation.

Economists may wonder why more household panel data have not
been collected in the past. The answer is that it is expensive
to establish and maintain a panel and to process panel data.
For a panel of 10,000 households, even at a cost of $50 per
household for recruitment and training, a low figure panel
establishment field cost represents a $0.5 million investment.
Compensation to reporting households at $50 per year entails
another $0.5 million. From a 75 percent reporting rate, 390,000
diaries would be generated to be edited, coded, processed, and
summarized. Postage to and from for each diary must be included. Operating costs could easily range between one and two
million dollars per year. Therefore, household panel data are
very costly inputs to a demand analysis.

Store Panels

The concept of continuous data from a sample of retail
stores was developed because of the high cost of household
panels and because household demographics are not essential to
all demand analyses. Increased accuracy usually occurs because
a single supermarket, on the average, reflects purchases by
between 1,000 and 1,500 households. Since sales coverage is
larger per single data generating point, the survey sample size
can be smaller, or else data accuracy may be greatly increased.
A long-standing private store audit data system is that of A. C.
Nielson.

A. C. Nielson Service. The Nielson data service, which began in
1923, now operates retail store audits in 23 countries. The
U.S. Census of Distribution serves as a major guide in selecting

the store sample in the United States. Sample stratifications
are based on the following characteristics: store type (gro-
cery, combination, delicatessen, general), store size (super,
large, medium, small), service type (self-service, clerk service), ownership (chain, voluntary chain, independent), income
area (upper, middle, lower).

The Nielson United States store sample comprises approximately 1,300 retail outlets which are audited every 60 days.
Sales audits cover only a 30-day period; therefore, stores'
auditing schedules are time staggered over the 60-day interval.
Product sales coverage is claimed to be equivalent to purchases
by 1.5 million households. Food sales audit data and a drug
product series are reported. Audits measure retail sales by
determining the quantity of products delivered to the store and
then adjusting those figures for beginning and ending store
inventories. Resulting data are net sales per producer per
store per time period by brand, package size, and average retail
price per unit.

Nielson clients utilize the data to evaluate product and
brand marketing sales positions by trade territories and to
measure the effects of company marketing strategies. However,
the data are useful for specialized types of demand analyses.
Data are compiled for the following twelve categories: (1)
sales to consumers, (2) purchase by retailers, (3) day's supply
on hand, (4) store count distribution, (5) all-commodity distribution, (6) out-of-stock occurrence rate, (7) prices (wholesale
and retail), (8) special factory packs, (9) dealers support
(displays, advertising coupon redemptions), (10) special observation (order size, reorders, direct versus wholesale channels),
(11) total food store sales (all commodities), and (12) major
media advertising (from other sources).

These data sets are summarized by geographic region, by 38
major SMSA markets, and by population size of counties (large
metro areas, other counties over 100,000 population, counties of
30,000 to 100,000, and rural--under 30,000.

In concept as well as in practice, food demand behavior
differs in the short run and long run. Short-run demand typically is more elastic with price. Yet, product marketing
occurs in the short run: daily, weekly, and monthly. It can be
argued that inadequate attention has been paid by economists to
short-run food demand analyses, but the problem mostly reflects
the lack of availability of the continuous data series from
either store audits or household panels because these have been
held in the private sector. Even when short-run period data are
available, economists usually have obtained long-run demand
parameters on the assumption that these are of interest and
concern to all government policy decisions.

Household Surveys

Because continuous consumer panel data for the United States
before 1980 were all in the hands of private research firms,

cross-sectional analyses before that time relied on data from periodic household surveys. Household surveys normally obtain food purchases, or consumption, for no more than a one- to four-week period because of survey cost escalations beyond that time dimension. Weekly records are the typical format. Purchase diaries and/or purchase recall is used. Two household surveys with national coverage are the well-known BLS Consumer Expenditure Surveys and the USDA Food Consumption Surveys.

Consumer expenditure survey. The BLS Consumer Expenditure Surveys (CES) are the responsibility of the Bureau of Labor Statistics which also maintains the Consumer Price Index (CPI). The two are interrelated. The CPI measures retail price changes nationwide for consumer products and services. In comparison, the central purpose of CES is to obtain expenditure information on various products and services so that the relative importance of each is properly reflected in the CPI. Prior to 1979, the consumer expenditure surveys were made about every ten years. During 1979, continuous CES were initiated with provision for summary reports on a quarterly basis. A prearranged rotation of households into and out of the panel differentiates it from a true continuing panel like that of MRCA or NPDR where replacements occur only as dropouts occur.

The 1972-73 CES was the last periodic survey preceding the continuing series. Like its predecessors, the surveys consisted of two parts--an interview recall survey and a diary record survey. The interview survey covered three months and was rotated across the sample of households during a two-year period. The diary survey covered two one-week periods. Purchase records of infrequent large price items were obtained from the recall survey phase because purchase incidence is too low on a weekly basis. As a general rule, diaries are more accurate for small item purchases. Recall is better for large purchases like semidurable or durable product items.

Sampling was based upon the 216 U.S. geographic subareas designed by the Bureau of the Census as its sample for the current population surveys. The diary phase sample totaled about 20,000 households or about 10,000 during each of the two years. Approximately 70,000 households comprised the interview survey since a larger sample is required for comparable accuracy regarding infrequently purchased items.

Demographic data for the CES include geographic region of the household; its urbanization category; occupation of household head; family size; age, education, and race of family head; number of income earners; and housing tenure. In addition to data given in published reports, more detailed information is contained on computer tapes and is available from the BLS. One key limitation of the CES to demand analysts is that price information is not included. Prices must be interpreted from the quantity and total expenditure figures, a procedure lacking the precision economists prefer because memory or recording

errors become embedded in the price variable. The same short-
coming applies to the USDA Household Food Consumption Survey to
which we now turn.

Household food consumption survey. As an information guide to
the formation of national food policy and nutrition education,
the U.S. Department of Agriculture conducts a national survey of
household food consumption approximately every ten years. The
first survey began in 1936. The most recent one was conducted
in 1977.

The sample size usually ranges between about 7,500 to 15,000
households depending upon the survey component involved. Data
for purchased and nonpurchased food are obtained. The latter
includes food from home gardens, farms, or gifts. A seven-day
recall methodology is employed for both quantities and expenditure figures. Prices, as noted above, must be inferred.

Menu studies are included using recall and/or diary information. Parts of the survey data rest on actual consumption and
parts on interpretations from food quantities brought into the
household. In the latter, allowances are made for food preparation and cooking losses. A major advantage of the Household
Food Consumption Survey is the greater individual food product
detail available as compared with the Consumer Expenditure
Survey. In the CES, for example, data for all canned citrus
juices and/or canned vegetable juices are combined rather than
being presented separately for each product. In addition,
household socioeconomic and demographic data are included.

Channel Movement Data
Another source of data besides store audits and household
surveys is measurement of product movement within the marketing
channels. Channel movement data are usually at the processor or
wholesale level. Like the other data systems previously discussed, these may be continuous or periodic. One continuous
private system is SAMI, an acronym for Statistical Area Market
Information. The SAMI data service was created by the publish-
ers of Time and Life magazines to meet the market information
needs of individual marketing firms. Its initial purpose was to
provide product sales data to the magazines' current or prospec-
tive advertising clients. SAMI records product shipments from
distribution warehouses to retail stores.

Totaled shipments are given for the last 52 weeks and are
compared on a 4-week basis to the same 4 weeks a year ago. Most
food chains in major SMSA markets are covered and most products
are included except for fresh meats, fruits, and vegetables.
Total quantities and their retail values by individual brand and
package size are reported in the survey. Prices must be derived, but in this case they are accurate.

Food chain warehouses typically service stores in outlying
cities as far as 100 miles or more away. Importantly, SAMI data
are adjusted to exclude shipments to outlying stores; therefore,

the data relate specifically to the specified SMSA area. SAMI, which began in 1966, now covers approximately 48 SMSA markets. Since SMSA areas grow in population over time, for demand analysis purposes, these data must be converted to a per capita or household equivalent basis.

Economic Censuses
    On the public information side, data are provided on a continuous basis by the Department of Commerce through the Survey of Current Business. The U.S. Bureau of the Census reports economic data useful to food demand analysis. Conducted about every five years, the census data are supplemented by a number of continuing intercensal reports. For example, those of particular interest to demand analysis are (1) 1977 Census of Retail Trade, (2) 1977 Census of Wholesale Trade, and (3) 1977 Census of Manufactures.
    Utilization of census data depends upon an adequate knowledge of the standard industrial classification (SIC) codes. SIC codes specify the business categories into which the manufacturing, wholesale, retail, or service level data are classified. A four-digit code denotes the industry and product category. If one decodes the SIC 5142 code for frozen food wholesaling, the 51 denotes "nondurable goods," the 4 indicates "grocery and related products," and the 2 signifies "frozen goods." Some data subclasses are provided beyond the four-digit code. For example, meat markets are distinguished from seafood markets though both bear code 5423.
    Census of Business data, corrected for inventory changes, show market equilibrium quantities at the existing price level for a product category. Product definitions should be consulted to be sure what is being measured. Usually, data for more product classifications are provided at the processing level than at wholesale or retail. To illustrate, five classes of dairy products are reported at the processors level, but only a "total dairy products" figure appears in the retail store data.
    Analysis of food sales on an individual state or SMSA market basis may require U.S. Census figures on food store sales. Since total sales now include a large component of nonfood items, some assistance is provided in separating out nonfood sales by utilizing the census Merchandise Line reports. In these, food and nonfood sales categories are listed separately.
    Census data can be supplemented by information appearing in the Current Business Reports series of the Census Bureau issued through the U.S. Department of Commerce. These provide monthly information on prices, levels of production, consumer income, inventory levels, exports, and imports. Although census and intercensal data are mostly useful for those models based on aggregate data, one should remember that the data can also serve as a check on the reliability (extent of bias) of data from other sources. However, to believe that census data always are the ultimate measure is an unwarranted assumption.

New Data Systems
     Shortcomings evident among the foregoing data systems have
led to other innovative efforts. One of the major problems is
the overwhelming tendency for time-series data to lack both
product and price details. In comparison available detailed
data from the periodic national consumer expenditure and food
consumption studies have lacked time continuity. Therefore,
three new data systems began emerging in the 1980s toward a
resolution of these problems. One is the continuing BLS Consumer Expenditure Survey. Another is the continuing USDA
Household Food Consumption Survey. Both are public-based data
sources. The third involves retail store scanner consumer
panels within the private data arena.

Continuing household surveys. Extension of the BLS Consumer
Expenditure Surveys into a continuing quarterly survey is a
landmark decision for consumer demand analysis economists. For
the first time in the United States, a continuing data series
with cross-sectional demographics is available from a public
source. Unfortunately, direct price information is still
missing. Also, quantity information so far is not published
although it is on the initial data tapes. Data from the original tapes should permit resurrection of most of the inferred
price information. In the early stages of the continuing BLS
CES, information obtained from the panel households is not as
detailed as it should be. For example, if a reported purchase
is simply listed as beef, an assumed allocation is made among
steaks, roast, and ground beef. Prices collected for the
Consumer Price Index have limited product coverage which cannot
match the detail of the Quarterly Household Survey.
     The continuing USDA Quarterly Household Food Consumption
Survey design provides for replacement of 2,000 families out of
the total of 4,800 once a year, thereby creating a continuous
rotating panel of households. Such a rotation can contribute to
additional statistical variance in the data. On the other hand,
researchers postulate that households agreeing to continuous
participation in a panel are likely not to be a truly representative sample of consumers. Therefore, only actual experience
will determine whether the BLS rotating sample is a superior
methodology.
     The scope, sample, and methodology of the continuous food
consumption survey which USDA plans to launch in 1986 are still
in the formative stages.

Retail store scanner data. The electronic scanner system was
developed to increase efficiency of customer check-out time at
food stores. Each package item carries a printed UPC (Universal
Product Code). The code has two five-digit segments coded into
bar symbols. The first five identify the manufacturer. The
second five denote the product, package size, flavor, color, and
other related information.

Scanners offer advantages to consumer panels. When a panel member shops at a scanner store, the cash register receipt lists products by name and purchase price. Thereby, recall error could be reduced for entries in consumer purchase diaries. A second and potential advantage is the ability to include a consumer identification code on store sales tapes. One can imagine a panel of households shopping at scanner stores. Purchase data could be recorded on the computer tapes fed by scanner systems making gathering of information less demanding on panel members. Such systems have been used by private data research companies in selected tests markets in which all major food stores have scanners. Members carry codes that identify their individual purchases. In addition socioeconomic and demographic data for each panel family can be directly associated with its food purchase record.

The richness of scanner data lies in the great detail in which it is available. The corollary is the cost and time-consuming effort necessary to reduce the mass of data to useful summary figures for demand analyses. Each week as few as 10 to 20 supermarkets will generate the equivalent amount of data of 10,000 household panelists. Elimination of nonfood items becomes part of the data management problem. The 1980s, it appears, will be learning years for scanner data assembly, management, and analysis.

Attitudinal and innovation adoption. Except for special market research and/or nutritional studies, the relationship between consumer food purchase behavior and psychometric consumer classifications is ordinarily not considered. Psychometrics is a term implying the classification of consumers as to their goal-drive or motivations which are based on beliefs and attitudes. Among the array of consumer classifications are (1) hedonists (junk food eaters), who only care about food tasting good; (2) dieters, who avoid or minimize sweets, fats, carbohydrates, or some other type of foods; (3) nutrition-conscious persons (including "natural food" advocates), who are especially sensitive to particular food additives; and (4) the moderates, who apply general rationalities to food buying without primary emphasis upon any particular viewpoint.

Consumer motivation, such as the foregoing, clearly affects food purchase behavior. Yet food demand analysts have generally ignored methodologies in demand models such as those formulated by brand product marketing researchers. Economists instead have created models based predominately upon the precepts of the Marshallian "rational consumer" model. A marriage of the basic demand theories and research concepts of economists, sociologists, and psychologists has been slow because professionals within the disciplines are largely unfamiliar with the concepts within the others. Challenged, thereby, are all food demand analysts concerned with food demand econometric model development.

NOTES

1. Contributing authors of this chapter in alphabetic order
are Barry W. Bobst, Robert E. Branson, Richard C. Haidacher, Eva
E. Jacobs, Robert Raunikar, Benjamin J. Senauer, David
Smallwood, Daniel S. Tilley, and Lillian R. de Zapata.
2. Additional information can be obtained from NPD Research
Inc., 15 Verbena Avenue, Floral Park, New York 10101; Tel.:
516-328-3941

REFERENCES

Folwell, R. J., and J. L. Baritelle. 1978. The U.S. wine market.
    USDA/ESCS, AER-417.
Gallo, A. E., J. Zellner, and D. Smallwood. 1980. The rich, the
    poor, and the money they spend for food. USDA NFR-11.
Hagar, C. J. 1979. Factors affecting fiber consumption in
    household textiles. USDA/ESCS Staff Rep.
Levine, D. B., and H. P. Miller. 1957. Response variation en-
    countered with different questionnaire forms. USDA/AMS
    Mark. Res. Rep. No. 163.
Purcell, J.C., et al. 1957. The Atlanta consumer panel. Ga.
    Agric. Exp. Stn. Mimeo Ser. N.S. 44.
Raunikar, R. 1976. The Griffin consumer research panel: Estab-
    lishment and characteristics. Ga. Agric. Exp. Stn. Res.
    Rep. No. 232.
Salathe, L. E. 1978. A comparison of alternative functional
    forms for estimating household Engel curves. Pap. at
    meetings of Am. Assoc. of Agric. Econ., Virginia Polytechnic
    Inst. and State Univ.
_____. 1979. Household expenditure pattern in the U.S. USDA/
    ESCS Tech. Bull. No. 1603.
_____, and R.C. Buse. 1979. Household food consumption pat-
    terns in the United States. USDA/ESCS Tech. Bull. No. 1587.
Smallwood, D. 1980. Analysis of food away from home food expen-
    ditures. USDA/ESCS Food Demand Res. Pap. No. 17.
Tyrrell, T. J. 1979. An application of the multinomial logit
    model to predicting patterns of food and other household
    expenditures in the Northeastern United States. Ph.D.
    diss., Cornell Univ.
U.S. Department of Agriculture. 1972. Major statistical series
    of the U.S. Department of Agriculture. Agriculture Handbook
    No. 365, vols. 1-11. Washington, D.C.: U.S. Government
    Printing Office.
_____. 1982. Agricultural prices, annual summary 1982. Crop
    Reporting Board, Stat. Rep. Serv. Washington, D.C.: U.S.
    Government Printing Office.
_____. 1982. Agricultural statistics 1982. Washington, D.C.:
    U.S. Government Printing Office.
_____. 1984. Food consumption, prices, and expenditures.

Stat. Bull. No. 713. Washington, D.C.: U.S. Government
Printing Office.
_____. 1985. *Cotton market news*. Memphis: Agricultural Mar-
keting Service - Cotton Division.
U.S. Bureau of the Census. 1979. *1977 census of wholesale
trade*. Washington, D.C.: U.S. Government Printing Office.
_____. 1979. *1977 census of manufactures*. Washington, D.C.:
U.S. Government Printing Office.
_____. 1979. *1977 census of retail trade*. Washington, D.C.:
U.S. Government Printing Office.
_____. 1981. *1978 census of agriculture*. Bureau of the Cen-
sus. Washington, D.C.: U.S. Government Printing Office.
Wold, H., and L. Jureen. 1953. *Demand analysis: A study in
econometrics*. New York: Wiley.
Young, T. 1977. An approach to commodity grouping in demand
analysis. *J. Agric. Econ.* 28:141-51.

CHAPTER 3

Data Problems in Demand Analyses: Two Examples

SOURCES OF DATA FOR DEMAND ANALYSES were discussed in the previous chapter. It was noted that although these data may accurately measure what was intended, their suitability for specific econometric models for consumer demand analysis is often inadequate. Furthermore, because data collections and maintenance of data bases are costly, they frequently represent a compromise. Problems related to the conceptualization, design, and operation of data systems are explored further before we consider two major data sources.

Data for demand analysis use should meet an extensive list of criteria that may be categorized as follows: (1) clear designation of subject to be measured, (2) stipulation of data accuracy targets, and (3) compliance with conceptual corollaries related to price index theory.

Identification of the subject to be measured depends on demand model specification. Therefore, some key considerations deserve mention and comment on their implications. Neoclassical theory pertains to behavior of a typical rational consumer--a microview--and, hence, if the assumptions are to be fulfilled, data inputs (e.g., price and purchase information) should be at the individual consumer level on a continuing basis.

Because of the foregoing type of differences between conceptual data standards and those of available data and for reasons noted in Chapter 2, data series should not be accepted without careful investigation of their definition and the methodology guiding their procurement and calculation. Keeping this warning in mind, attention is now given to two major data sets which can exemplify the nature and extent of the problems. In what follows, discussion on specific data problems in demand analyses are focused on the widely used disappearance data system and the consumer expenditure surveys.

MEASUREMENT PROBLEMS IN DISAPPEARANCE DATA SYSTEMS
    Three problem areas are of special concern regarding disap-
pearance data systems.[1] The first is the probable existence of
measurement error. Second is the effect such error may have on
conventional commodity market analyses. And third is the interpretation problem when disappearance data are from markets in
disequilibrium.

Extent of Measurement Error

    The residual nature of disappearance estimates of demand was
mentioned in Chapter 2. The following equation (3.1) illustrates
the estimation procedure:

$$\text{Disappearance}_t = \text{production}_t + \text{change in observable stocks}_t$$
$$+ \text{ imports}_t - \text{exports}_t - \text{nonfood consumption}_t$$
$$- \text{nonmarket food consumption}_t \qquad (3.1)$$

It was noted that if the market in question is in equilibrium
and all the variables on the right side of the equation are
measured without error, then disappearance in period t is
conceptually equivalent to the quantity demanded in that period,
which is the quality of data needed for demand analyses.
However, many possible random measurement errors and bias
potentials exist in disappearance data. Data inputs for any or
all variables could contain these errors. Errors in different
data sets may be correlated rather than random. Given these
possibilities, it helps to make two simplifying assumptions
about the data reporting systems and procedures in order to
focus on the effects of random measurement errors on demand
analysis.
    The first and most important assumption relates to bias in
measurement. It is assumed that the organizations providing
data components for the disappearance estimates are operating on
a best-efforts basis and, therefore, these data are unbiased.
The second important assumption is that any measurement errors
are independent and uncorrelated. This independence assumption
is made more likely because of the independence of the agencies
providing the data components. The Economic and Statistical
Service (ESS) of USDA is responsible for making the
disappearance estimates. The Crop Reporting Service of USDA is
responsible for the production data component, the major figure
in calculating disappearance estimates. Fortunately, these two
USDA agencies operate with considerable independence. Bureau of
the Census, U.S. Department of Commerce, provides monthly import
and export data. Nonmarket food consumption data originate from
the Department of Defense (military procurement) and, where
applicable, from the Food and Nutrition Service of USDA for USDA
food donations. The Internal Revenue Service provides data on

grain utilization in beer and liquors.

Unbiasedness of error components implies that the expected value of measurement error in disappearance estimates is the sum of the expected values of its error components, viz.,

$$E(V_d) = \Sigma E(e_i) = 0 \tag{3.2}$$

where V_d = measurement error in disappearance and e_i = measurement error in the ith component of disappearance. The assumption of independence of error components implies that the variance of measurement error in disappearance data will equal the sum of the variances of its components, viz.,

$$E(\sigma_{V_d}^2) = \Sigma E(e_i^2) = \Sigma \sigma_i^2, \ E(e_i, e_j) = 0 \tag{3.3}$$

The foregoing assumptions hold unless other significant measurement error is found in the demand data.

Magnitudes of Measurement Error

Determining measurement error is difficult as indicated by the work of Morgenstern. Morgenstern (1963, 202-15) evaluated agricultural data by comparing census and USDA data for comparable years. He also evaluated the patterns of revisions in USDA data. His procedure is followed here for census year 1974. Production data are examined since measurement errors of production, as previously noted, can be a major error source in disappearance estimates.

Production estimates from the 1974 Census of Agriculture and from USDA for selected crops are displayed in Table 3.1. Census data for "all farms" and USDA's preliminary and revised estimates and percentage differences between them are noted in a format comparable with Morgenstern's table (pp. 202-03). Several comparisons can be made between Morgenstern's tables for 1949 and 1954 and those for 1974. First, discrepancies between census and USDA data are entirely one-sided, census estimates being consistently lower than USDA. This finding was generally the case both in 1949 and 1954. Magnitudes of the discrepancies were approximately the same as in the past, although USDA made smaller revisions in its series for 1974 than in the previous years.

The data in Table 3.1 do not measure measurement error, but they do imply that it exists, even though at first glance the one-sidedness of the discrepancies suggest that bias is the more important problem in these production series. However, both agencies can be presumed to be statistically competent, so the fact that neither agency adjusts its series to conform to the other's suggests that each considers its methods and results to be at least as reliable as the other. When competent organizations disagree in this manner, it suggests that inherent

TABLE 3.1. Comparison of census and USDA production data on selected crops, 1974

		Production			Percentage differences		
Crop[a]	Unit	(1) Census,[b] all farms (1,000 units)	(2) Preliminary,[c] USDA (1,000 units)	(3) Final,[d] USDA (1,000 units)	(4) = (1)-(2)/(1)	(5) = (1)-(3)/(1)	(6) = (2)-(3)/(2)
Corn	bu	4,396,913	4,651,167	4,701,402	− 5.8	− 6.9	−1.1
Sorghum	bu	554,244	628,081	622,711	−13.3	−12.4	0.8
Wheat	bu	1,691,553	1,793,322	1,781,918	− 6.0	− 5.3	0.6
Soybeans	bu	1,145,788	1,233,425	1,216,287	− 7.6	− 6.2	1.4
Cotton	bales	10,887	11,540	11,540	− 6.0	− 6.0	0.0
Peanuts	lb	3,168,918	3,667,604	3,667,604	−15.7	−15.7	0.0
Potatoes	cwt	316,164	340,116	342,395	− 7.6	− 8.3	−0.7

[a]Production measures exclude production for silage on forage use.
[b]U.S. Department of Commerce, 1974 Census of Agriculture (1977).
[c]U.S. Department of Agriculture, Agricultural Statistics 1975 (1975).
[d]U.S. Department of Agriculture, Agricultural Statistics 1979 (1979).

measurement errors are present in both series. Thus, it is important to consider the effects of data measurement error on time-series demand analysis.

Effect of Measurement Error

Evaluation of the impact of measurement error on demand estimation is a part of the important errors-in-variables problem. Although it has been implied in econometric literature (Theil 1971, 614), it has not been spelled out in any detail. It is worthwhile to do so here.

The single equation, static model case. Consider the following demand function for a commodity.

$$QD_t = GP_t + BX_t + UD_t \tag{3.4}$$

where QD_t = quantity demanded, P_t = price of the commodity, X_t = set of exogenous variables affecting demand, UD_t = disturbance term, and G,B = parameters. The disturbance term has the usual interpretation of referring to diffused economic events that affect demand in a random manner. It is distinct from measurement error that might exist in the data series for QD. Assume that P is predetermined and that it and the variables in X are uncorrelated with UD. Assume also that UD meets the standard conditions (zero expected value, homoscedasticity, and serial independence) necessary for efficient least-squares regression. If QD is observed without error, then an unbiased estimation of the model's parameters is obtained.

Let us now consider the effects of measurement errors in the data for QD. The model can be rewritten as

$$QD_t = GP_t + BX_t + WD_t, \qquad WD_t = UD_t + VD_t \tag{3.5}$$

where VD_t = measurement error, and WD_t = sum of disturbance and

error terms. The combined disturbance-error term WD retains the
properties of its components, which is to say that it behaves as
a random variable that is uncorrelated with the regressors P and
X. If this is the case, then the necessary conditions for least-
squares regression are preserved and consistent parameter estimates can still be obtained. To be sure, the estimates will be
less precise than if QD were measured without error. The expected value of the variance of the estimate is now

$$\sigma_{WD}^2 = \sigma_{UD}^2 + \sigma_{VD}^2 \tag{3.6}$$

instead of σ_{UD}^2, and standard errors of the parameter estimates
will be increased in proportion to the magnitude of σ_{VD}^2.

The static systems of equations case. Since commodity demand is
more often analyzed in the context of a system of equations
specifying said commodity's market than as a single equation,
the effect of measurement error on parameter estimation in such
models needs to be appraised. Consider the model

Demand: $QD_t = G_{11}P_t + B_1 X1_t + WD_t$

Supply: $QS_t = G_{21}P_t + B_2 X2_t + WS_t$

Stocks change: $K_t = G_{31}P_t + G_{32}QD_t + B_3 X3_t + WK_t$

Market clearing: $K_t = QS_t - QD_t$ \hfill (3.7)

where QS = quantity supplied; K = period to period changes in
stocks; X1, X2, X3 = vectors of exogenous variables affecting
demand, supply, and stocks changes, and WS, WK = combined
disturbance-measurement error terms for the supply and stocks
change functions.

   It is assumed that X1, X2, and X3 contain a sufficient
number of variables to overidentify the model. It is assumed
from the outset that data series for all the quantity variables,
QD, QS, and K contain measurement errors. Price is no longer
assumed to be predetermined, instead it is jointly determined
with the commodity's quantity variables. It is assumed to be
measured without error.

   It is well known that measurement errors in regressor
variables cause bias in least-squares regression estimates of
parameters. The extraneous variation from measurement errors
cause parameters estimates to be attenuated from their true
values; that is, they are biased towards zero (Garber and
Klepper 1980). In view of the inclusion of QD on the right side
of the stocks change function in (3.7) it might be thought that
least-squares   bias   must   affect   parameter   estimates.
Paradoxically, however, it is the complexity of the estimation
methods   used   for   systems   of   simultaneous   equations   that

simplifies the matter so far as least-squares bias is concerned, at least in static models.

This correction for least-squares bias is a side effect of the stagewise approach used by the parameter estimation methods appropriate to systems of simultaneous equations. It has long been established that endogenous variables which are included on the right sides of equations in simultaneous systems, such as P in the demand and supply functions and both P and QD in the stock change function of (3.7), will be correlated with the disturbances of those equations. As a consequence, least-squares regression is an inappropriate estimation technique. The stagewise regression procedures, such as two-stage or three-stage least-squares, purge the included endogenous variables of their correlations with disturbances by replacing them with predicted values generated from first-stage regressions. In the case of the stocks change function, the variable QD would be replaced by QD*, which is estimated by

$$QD_t^* = A_0 + A_1 Z_t \qquad\qquad (3.8)$$

The coefficient vectors A_0 and A_1 of equation (3.8) are estimated by least-squares regression of QD on the exogenous variables in Z, which are the variables in X1, X2, and X3 after duplicates are eliminated. Unbiased estimates of these coefficients can be obtained despite measurement error in QD because QD is treated as a dependent variable in the regression. Replacement of QD by QD* in second and later stages of the simultaneous equations estimation procedures not only purges the disturbance term, it also purges any measurement error that might have been present.

All endogenous variables appearing as regressors in a simultaneous equations model are replaced by first-stage least-squares estimates, so their measurement errors will not affect parameter estimates obtained in the later stages of the procedures. Errors remain in the endogenous variables when they are used as dependent variables in these stages, but they do not bias the parameter estimates for the same reasons that errors in dependent variables do not affect single equations regression estimates. Also, it can now be seen that it was unnecessary to assume that P was measured without error. Consistent parameter estimates can be obtained for static, simultaneous models despite measurement errors in any or all of the endogenous variables.

Pseudo-correlation between equations and choice of estimator. Measurement error will cause some correlation between the disturbance terms of demand and other quantity-dependent equations even though their disturbance terms might be independent. This result is due to the residual nature of observations of quantity demanded (disappearance). For example, suppose K in model (3.7) is observed without error. In that case, all the measurement errors in QS will be reflected in QD

so that the measurement error components of the demand and
supply disturbance terms will be perfectly correlated. The
resulting correlation between the disturbance terms WD and WS
will lie between 0 and 1 depending on the relative weights of
disturbances and errors.

Three-stage least-squares provides asymptotically more
efficient parameter estimates than two-stage least-squares when
correlation exists between equation disturbances. Since this
will invariably be the case when measurement error exists in the
quantity data, three-stage least-squares estimation will tend to
be preferred. However, other factors, such as the added chance
of specification error, may outweigh this one in choosing an
estimation method.

Measurement error-induced bias in dynamic models. Demand is
often hypothesized to be affected by past events in the manner
of distributed lags. Consider the model

Demand:

$$QD_t = G_{11}P_t + H_1 QD_{t-1} + B_1 X1_t + WD_t$$

Supply:

$$QS_t = G_{21}P_t + B_2 X2_t + WS_t$$

Stocks change:

$$K_t = G_{31}P_t + G_{32}QD_t \ B_3 X3_t + WK_t$$

Market clearing:

$$K_t = QS_t - QD_t \tag{3.9}$$

which is different from (3.7) by the addition of variable QD_{t-1}
to the specification of demand. This specification is the estimating form of the common Nerlove hypothesis of incomplete
adjustment of demand to its equilibrium level within the time
period of observation. If QD_t contains measurement errors,
the QD_{t-1} does also and parameter estimates will be biased as a
result.

Origins of these biases in parameter estimates are twofold.
The first source is the inclusion of lagged endogenous variables
containing measurement errors in the first-stage estimates of
current endogenous variables. For the model described in (3.9)
these estimates are

$$QD'_t = F'_1(Z_t QD_{t-1})$$

$$QS'_t = F'_2(Z_t QD_{t-1})$$

$$K'_t = F'_3(Z_t QD_{t-1})$$

$$P'_t = F'_4(Z_t QD_{t-1}) \tag{3.10}$$

where F'_1, F'_2, F'_3, and F'_4, are the vectors of least-squares regression coefficients estimated on the right-hand variable matrix, which includes QD_{t-1}. Measurement error in QD_{t-1} causes attenuation of its regression coefficients in $F_1' - F_4'$ which in turn leads to inconsistent first-stage estimates of the endogenous variables. For positive coefficients on QD_{t-1}, the bias will cause the endogenous variable estimates in (3.10) to be underestimated when QD_{t-1} is large and overestimated when it is small. Therefore, the requirement of unbiased first-stage estimation in two-stage and three-stage least-squares is not met. Furthermore, the effect of the bias spreads to all the equations in model (3.9) rather than being confined to the one equation in which QD_{t-1} appears.

The second source of parameter bias is the use of QD_{t-1} as a regressor in the second or third stage of the estimation procedure. In two-stage least-squares this secondary bias effect is limited to the equations containing lagged endogenous variables with measurement errors. In the example model (3.9), secondary bias would be restricted to the demand equation, but it would tend to spread to all demand equation parameters in proportion to the amount of measurement error in QD_{t-1}. In three-stage least-squares, however, secondary bias spreads to all equations in the model through matrix inversions involved in the generalized least-squares technique that is used in the third stage.

There does not seem to be any practical way of modifying estimation techniques to avoid the biases caused by measurement errors. Several methods have been suggested in the context of ordinary least-squares, including Wald's grouping method, weighted regression, and instrumental variables (Malinvaud 1970, 383–400). Wald's method and weighted regression call for conditions or prior information requirements that are impractical to meet. The instrumental variables method seems promising at first because the first-stage estimates developed in two- and three-stage regression methods are fundamentally instrumental variables, so substituting instrumental variables for lagged endogenous variables would be an extension of the estimating method rather than an ad hoc addition. However, a straight extension of the first-stage technique implies that the estimator for QD_{t-1} in (3.9) is

$$QD_{t-1}^* = F_5(Z_{t-1}QD_{t-2}) \qquad (3.11)$$

which is the first-stage equation for QD_t in (3.10) lagged one period. Measurement error is merely lagged, not eliminated. Dropping QD_{t-2} from (3.11) would eliminate the measurement error but at the cost of committing a specification error. Aigner

(1974) has shown that eliminating an error-ridden variable
usually is more costly than retaining it in the specification of
single-equation regressions. This finding suggests that the
first-stage estimates

$$QD_t = F_6(X_t Y_t Z_t X_{t-1} Y_{t-1} Z_{t-1})$$

$$QS_t = F_7(X_t Y_t Z_t X_{t-1} Y_{t-1} Z_{t-1})$$

$$K_t = F_8(X_t Y_t Z_t X_{t-1} Y_{t-1} Z_{t-1})$$

$$P_t = F_9(X_t Y_t Z_t X_{t-1} Y_{t-1} Z_{t-1}) \qquad\qquad (3.12)$$

would usually be an inferior alternative to the estimates in
(3.10). Besides, the regressions in (3.12) are obviously more
expensive in terms of degree of freedom and are subject to more
multicollinearity problems if there are strong trends in any of
the exogenous variables.

Obviously there is no substitute for good data in the estimation of dynamic commodity models. The effects of measurement
errors in lagged endogenous variables permeate all equations in
the model in one fashion or another. More specifically, static
demand equations do not escape the effects of measurement error
if they are included in a model containing a dynamic supply function, and dynamic supply functions are commonly used in commodity model specifications.

CONSUMER EXPENDITURE SURVEYS:
THEIR PROBLEMS, USES, AND ABUSES
    In the United States, BLS consumer expenditure surveys (CES)
are a very rich source of information on household expenditure
behavior.[2] In addition to expenditure data, they contain a very
wide array of information on household characteristics. Many,
if not all, of these variables affect the individual unit's economic behavior. Thus, expenditure models derived from cross-
sectional expenditure surveys will be much richer than those
derived from time-series in that they can utilize many more
variables and, hence, provide the analyst much more information. It is also generally accepted that cross-sectional data
provide more acceptable estimates of income elasticities than
those obtained from time-series data (Prais and Houthakker 1955,
227-29). In time-series data, price and income tend to be correlated and thus, the resulting estimates are difficult to interpret and use in hypothesis testing and policy analysis. Expenditure surveys yield more accurate estimates of the appropriate
elasticities for program analysis because they are considered to
be long-run elasticities (George and King 1971, 275). In view
of these attributes, CES data are likely to be increasingly used
in demand analyses.

Three consumer expenditure surveys have appeared in the past 20 years. There are many hidden data reefs and theoretical assumptions that can impair a researcher's model when using CES data. The following discussion concerns ideas and concepts that need clarification if one is to successfully utilize these surveys. It is important to evaluate these data bases critically and carefully in relation to their use in demand analyses.

Efficient use of large data bases requires sophisticated data manipulation and organizational techniques. But few researchers have much experience with the statistical software and computing systems required to efficiently and systematically exploit the full potential of the data sets. Furthermore, if one obtains CES data sets in machine readable tape form, one is once removed from the actual data. The unquestioned acceptance of the data in this form is a tempting and easy trap.

Several model/data conflicts are the subject of the section which follows. They are discussed in terms of exigent areas of work.

Model/Data Conflicts

Model construction takes place in three phases, the economic model, the mathematical model, and the statistical model. All purport to be an abstract representation of observed behavior. King (1979) describes some of the conflicts that can arise among the three, leading the researcher into unwarranted or even completely erroneous conclusions. He does not discuss the problem the researcher encounters when he or she parameterizes a model with data that does not conform to model assumptions. This section extends that discussion to the consumer expenditure surveys that are increasingly being used to parameterize such models. There are two types of problems. One is a simple data problem arising from inconsistent, impossible, or improbable data for a particular consumer unit (CU). The other is the conflict between the implicit assumptions of the mathematical and statistical model and the data base. If not recognized by the researcher, both can lead to poor results and/or erroneous conclusions.

Data problems. By their very nature, household expenditure surveys on a national scale involve obtaining and recording massive numbers of pieces of information. The agency responsible for the survey and the production of the public use tapes (PU-tapes) diligently tries to locate and reconcile errors and inconsistencies. Generally, a very creditable job is done. Nevertheless, more than a few errors remain. One survey staff group has difficulty in examining data from all technical points of view. Improved communication from researchers, as the data users, can help address and correct the problems. Problems encountered in the 1972-73 and 1973-74 BLS PU-tapes for the Diary Survey (CEDS) and Interview Survey (CEIS) emphasize the need for considerable care and judgment in using the tapes.[3]

Keep in mind that the data were not generated with demand analysis criteria in mind.

Inconsistencies and other errors are summarized in Table 3.2. To determine these, a large number of tests were performed on the data for each of the consuming units included in the 1972-73 and 1973-74 BLS CEDS. Some cases reflect errors in translating data into machine readable format. In others, it was data misinterpretation or coding. Many cases were easily resolved by the application of simple logic. Others were not resolvable without returning to the original survey documents, but government guarantees of anonymity prevented that. BLS is most cooperative in resolving data problems, but budget restrictions limit the time and resources available for such matters.

TABLE 3.2. Number of consumer units in the 1972-73 and 1973-74 BLS CEDS PU tapes exhibiting data problems in socioeconomic information

Data problem	Diary year	
	1972-73	1973-74
Inconsistent start dates	20	51
Missing start dates	38	29
FM detail inconsistent with reported aggregates	77	108
Incomplete food stamp data	17	232
Weeks worked missing	28	--
Alpha-numeric data entry	11	25
Income detail does not sum to total	--	3
Total	191	448
Total consumer units	11,065	12,121
Percentage with errors	1.7	3.7

Source: Buse (1979).

Several conclusions emerge from Table 3.2. First, the detectable error rate in the socioeconomic information is quite small--less than 3 percent. Second, although the error rate is small, it can prove troublesome because errors may be outliers and contribute substantially to overall data statistical variances. Examples are zero dates which create expenditure time frame inaccuracies, zero expenditures for food stamps, mispunched diary records start dates, or negative earnings. Encountering alphabetical data where only numbers are expected cause potential expensive computer run failures. Other but similar kinds of data problems are noted in Table 3.3. Again, the error rate is not large, but it does illustrate that some problems can be troublesome. This set can be labeled as data errors (correctable) or data warnings (uncorrectable) because information is unavailable thereto. Inconsistencies also arise where both correctable error and warnings apply. In 30 cases, involving the presence of family members (FMs) in the CU, one or more family members over 1 year of age were reported as being in the CU zero weeks in the past year. Although the code for "weeks in the CU" would be changed from 0 to 99 by the analyst,

TABLE 3.3. Number of consumer units in the 1972–73 BLS CEDS exhibiting data
problems in socioeconomic data

Data problem	No. of CUs
Data errors	
Marital status inconsistent with reported data on FM-2	127
Weeks FM was a member of CU not reported	30
Average family size incorrectly calculated	3
Incorrect sign on earnings	20
Earnings of "other" set to zero because no FM detail present	20
Earnings of other was wrong sign	2
Data warnings	
Calculated family size inconsistent with FM-detail	27
Exchange value of food stamps less than cost	12
Exchange value of food stamps not reported	2
Cost of food stamps not reported	57
Earnings detail inconsistent with reported	6
Total	306
Total consumer units	19,975
Percentage with errors and warnings	1.53

the average family size could not be changed; only a data
warning could be incorporated into that CU record to warn the
user of a potential problem. In 127 cases, FM-1 was listed as
married but there was no adult male or female listed among the
FMs. In this case, the marital status of FM-1 was changed from
1 (married) to 2 (not married).

Data conflicts. A researcher faces two questions before analyses can proceed. First, what level of detail (aggregation)
produces the most useful results for public policy analysis?
Detailed data can always be aggregated but it increases the cost
exponentially. This trade-off leads to an alternate question:
What is the maximum level of aggregation that will not obscure
significant differences in the demand parameters? For example,
meat expenditure may be unsatisfactory in that it hides the behavioral differences between pork, beef, and poultry. At the
other extreme, such differences as may exist between chuck roast
and round steak, or hair shampoo and hair conditioner, may not
be worth the added analysis cost. The policy analyst can provide some limits for the first question. The alternate question, however, can only be answered through knowledge of the
underlying structure. This knowledge can only be obtained by
repeated empirical analysis, by theoretical guidelines, or by
both.

The second operational question facing the researcher is how
to handle the reporting of unusually large or small values of a
variable. Such outliers may cause problems. They could be hypothesized to have arisen out of alternative consumer purchase
behavior models or simply be erroneous data. In either case,

the researcher should examine any data set for the size and frequency of extreme values.

Given the nature of a consumer panel two-week reporting
period, it is axiomatic that the more detailed the level of
expenditure to be examined, the larger the proportion of households that will likely have zero values for a particular item.
But some analytical models and estimation procedures assume
positive quantities in the dependent variable observations.
Thus, attention to data density becomes a prerequisite to any
analysis.

Data aggregation. The theoretical literature provides little
guidance concerning how to aggregate, or group, expenditures
without obscuring the economic structure one is trying to
understand. The conservative approach would be to estimate
expenditure functions at the lowest level of disaggregation.
The data tape from the survey contains almost 2,300 expenditure
items and over 300 financial items. A particular CU will only
report some of the possible items. Obviously, the larger the
percentage of zero expenditures on a particular item, the smaller the total variation that must be explained. However, the
mathematical and statistical demand models assume that each CU
has positive nonzero values on the dependent variable, expenditure, but contrarily the data may contain a high proportion of
zero expenditures thereby creating a conflict between model
assumptions and the actual data. At present, there is no literature evaluating the impact of data density on either the
equation's parameter estimates or on the coefficient of multiple
determination. Nevertheless, the researcher must consider the
possible effect of numerous zero-value observations.

Data density. Data density is defined as the proportion of CUs
containing nonzero values for a particular expenditure. It is
useful to examine food expenditure data as well as that for other
categories of household expenditures. An overview for food will
be addressed first.

One analysis of the 1972-73 BLS CEDS data tapes for zero
observations considered information in 9 major categories of
expenditures.[5] Principal emphasis was on food expenditures.
The 212 different types of food items were aggregated into 12
subsets and into food at home and away from home. Eight nonfood
expenditures were examined. For 4 of the 9 major expenditure
classes shown, two-thirds or more of the households reported
expenditures (Table 3.4). For the other 5, between two-fifths
and two-thirds of the households reported expenditures during
the two-week diary period the tape covers. For these 9 expenditure categories, the researcher would have few density problems
even at the second level of disaggregation of food expenditures.

The same pattern exists in the 13 major expenditure categories of the 1972-73 BLS Survey (Table 3.5). Most have a density

TABLE 3.4. Density of major expenditure in the 1972-73 BLS CEDS: number of
consumer units reporting nonzero expenditure by category of expenditure

Expenditure category	CUs reporting nonzero expenditures	
	Number	Percent
Total food	22,233	95.9
Food at home	21,837	94.2
Cereal and bakery products	21,040	90.7
Meat, fish, and poultry	20,771	89.6
Dairy producs	20,943	90.3
Fruit	18,784	81.0
Vegetables	19,406	83.7
Sugar and sweets	14,922	64.4
Fats and oils	15,007	64.7
Nonalcoholic beverages	18,586	80.2
Miscellaneous foods	18,748	80.9
Food away from home	18,095	78.0
Meals	15,637	67.4
Snacks	11,527	49.7
Beverages	8,895	38.4
Alcholic beverages	9,741	42.0
Tobacco and smoking supplies	12,049	52.0
Personal care	16,388	70.7
Nonprescription medicines	9,690	41.8
Housekeeping supplies	19,150	82.6
Utilities and fuels	9,511	41.0
Automobile fuel and lubricants	17,070	73.6
Miscellaneous	14,058	60.6
Total consumer units	23,186	100.0

of 85 percent or more. Only 4 have a density below 66 percent.
The 4 may present special problems when using models of complete
demand systems.

As soon as one considers disaggregating below the level of
Tables 3.4 or 3.5, data density falls rapidly. For example,
89.6 percent of the completed diaries reported meat, fish, and
poultry expenditures (Table 3.4). However, data density falls

TABLE 3.5. Density of major expenditures in the 1972-73 BLS CEDS: number of
consumer units reporting nonzero expenditure by expenditure category

Expenditure category	CUs reporting nonzero expenditures	
	Number	Percent
Food	19,924	99.7
Alcohol	12,773	63.9
Tobacco	11,286	56.5
Housing	19,910	99.7
House furnishings and equipment	17,705	88.6
Clothing and material	19,734	98.8
Transportation	18,818	94.2
Medical care	19,237	96.3
Personal care	16,866	84.4
Recreation	18,161	90.9
Reading	16,835	84.3
Education	4,920	24.6
Other expenditures	13,016	65.2
Total consumer units	19,975	100.0

to between 66 and 72 percent when the meat category is disaggregated into beef and veal and pork (Table 3.6). Still further disaggregation drops it to between 4 to 47 percent. Recently, Salathe used the 1972–73 BLS CEDS to estimate the income and household size elasticities of 117 household expenditures of which 109 were food items. When 84 percent of the CUs report no expenditures, the question arises of how much reliance can be placed on an income elasticity for chuck roast of 0.247 based upon 16 percent of the CUs reported expenditures on that item. It clearly becomes a question of whether one uses a model comprising all households or treats each food item separately. In the latter case, one treats those CUs with expenditures as a subsample with the assumption that they are a representative sample.

TABLE 3.6. Density of selected subaggregates of household food expenditure in the 1972–73 BLS CEDS

| | CUs reporting nonzero expenditures | |
Expenditure category	Number	Percent
Meat, fish, and poultry	20,771	89.6
Beef and veal	16,682	71.9
Ground beef	10,869	46.9
Chuck roast	9,767	16.2
Round steak	2,492	10.7
Other beef	3,913	16.9
Veal	823	3.5
Pork–except canned	15,331	66.1
Bacon	8,483	36.6
Chops	6,014	25.9
Sausage	5,882	25.4
Roasts	1,302	5.6
Other	3,909	16.9
Total consumer units	23,186	100.0

A similar pattern emerges if other survey expenditures are disaggregated. For example, 99.7 percent of the CUs reported housing expenditures (Table 3.5). However, when one attempts to disaggregate into subcategories such as shelter, household operation, utilities, and telephone, the density falls to 55 percent for shelter compared to around 90 percent for each of the other three subexpenditures.

Detailed examination of the data suggests that most CUs with suspect data report few expenditures in one to three categories such as food, housing, and medical care. Thus, one strategy might be to eliminate CUs with less than 3 or 4 expenditures. Respondents grouped according to the number of nonzero expenditures reported are presented in Table 3.7. Here the researcher quickly faces a density problem, particularly in the diary based data. Even omitting households reporting 2 or fewer expenditures would eliminate 2,170 CUs or almost 10 percent of the sample.

A higher level of aggregation in the interview data reduces the severity of the density problem but does not eliminate it.

TABLE 3.7. Distribution of consumer units by number of major categories of nonzero expenditures reported, 1972-73 BLS CEDS and CEIS

Number of nonzero expenditures	Diary survey		Interview survey	
	Number	Percent	Number	Percent
0	812	3.5	0	--
1 or fewer	1,215	5.2	3	--
2 or fewer	2,170	9.4	18	0.1
3 or fewer	3,703	16.0	60	0.3
4 or fewer	6,092	26.3	170	0.9
5 or fewer	9,631	41.5	426	2.1
6 or fewer	14,242	61.4	810	4.1
7 or fewer	18,858	81.3	1,542	7.7
8 or fewer	22,061	95.1	2,672	13.4
9 or fewer	23,186	100.0	4,754	23.8
10 or fewer	--	--	8,431	42.2
11 or fewer	--	--	13,407	67.1
12 or fewer	--	--	18,203	91.1
13 or fewer	--	--	19,975	100.0

If the researcher wished to quantify a model for those house-holds reporting expenditures in the 6 expenditure categories of food, housing, clothing, medical care, transportation, and re-creation, there would be 85.5 percent of the survey CUs in the analysis (Table 3.8). However, as the expenditure categories increase from 6 to 9 by including house furnishing and equip-ments, reading, and personal care, the percentage of CUs reporting combinations of those expenditure declines to 64.4 percent.

To summarize, there is a great mass of detail in the expendi-ture surveys, a great need for more detailed models, and thus, a natural desire to parameterize detailed expenditure models. The rapid falloff in data density as one disaggregates raises doubts as to the validity and interpretation of such estimates.

Outliers. An outlier can be defined as an observation that lies outside of the general swarm of observations (an extreme value in some sense). Outlier values can seriously affect statistical results, particularly in ordinary least squares (OLS) estimates. Furthermore, the presence of outliers can indicate special cir-cumstances warranting further investigation of interactions, missing variables, and so forth.

TABLE 3.8. Number of consumer units reporting specific combinations of expenditures for 1972-73 BLS CEIS

Expenditure set[a]	Number	Percent
I	18,977	95.0
I, II	17,080	85.5
I to III	15,860	79.4
I to IV	12,855	64.4
I to V	4,761	23.8
All sets	1,772	8.9
Total consumer units	19,975	100.0

[a]Expenditure sets are: I = Food, housing, clothing, medical care; II = Transportion, recreation; III = House furnishings and equipment; IV = Reading, personal care; V = Alcohol, tobacco, other expenditures; VI = Education.

TABLE 3.9. Descriptive statistics on average biweekly expenditure of consumer
units reporting nonzero expenditures, 1972–73 BLS CEDS

	Biweekly expenditures in dollars			
Expenditure category	Mean	Standard deviation	Largest value	No. CUs > 5 S.D.[a]
Total food	69.04	58.00	4,099.93	45
Food at home	51.26	40.00	1,018.52	63
Cereal and bakery products	6.30	5.00	118.63	56
Meat, fish, and poultry	21.40	24.00	1,022.28	86
Dairy products	7.32	6.00	107.13	66
Fruit	3.98	5.00	285.20	34
Vegetables	4.45	4.00	76.33	64
Sugar and sweets	2.26	3.00	66.70	70
Fats and oils	2.05	2.00	56.00	64
Nonalcoholic beverages	4.40	4.00	117.35	48
Miscellaneous foods	4.86	5.00	231.54	50
Food away from home	22.96	40.00	4,000.00	30
Meals	21.94	41.00	4,000.00	25
Snacks	4.63	5.00	88.78	55
Beverages	2.14	5.00	355.65	25
Alcoholic beverages	10.76	15.00	434.76	49
Tobacco and smoking supplies	8.41	7.00	122.87	38
Personal care	8.20	11.00	731.50	51
Nonprescription medicines	5.74	14.00	518.97	73
Housekeeping supplies	6.40	8.00	267.40	87
Utilities and fuels	32.21	32.00	988.80	32
Automobile fuel and lubricants	19.01	20.00	1,245.78	65
Other expenditures	8.29	19.00	640.24	81
Total expenditure	125.18	92.00	4,109.52	39
Total annual income	11,657.95	13,823.00	6,650,000.00	

[a]S.D. = Standard deviation.

TABLE 3.10. Descriptive statistics on average annual expenditures of consumer
units reporting nonzero expenditures, 1972–73 BLS CEIS

	Annual expenditures in dollars			
Expenditure category	Mean	Standard deviation	Largest value	No. CUs > 5 S.D.[a]
Food	1774	1262	17,449	48
Alcohol	131	219	6,000	78
Tobacco	227	166	2,028	31
Housing	2129	1591	46,433	71
House furnishings and equipment	442	648	12,907	97
Clothing and material	627	660	11,463	85
Transportation	1912	2109	39,690	56
Medical care	493	578	31,239	67
Personal care	120	127	2,356	50
Recreation	339	450	21,596	64
Reading	58	75	1,581	101
Education	444	885	11,469	40
Other expenditures	126	393	26,417	68
Total expenditure	804	5201	99,716	38
Total annual income	11,658	13,823	6,650,000	

[a]S.D. = Standard deviation.

TABLE 3.11. Details on two largest outliers for each major expenditure
    category, 1972–73 BLS CEDS

Expenditure category	Amount (dollars)	Total expenditure (dollars)	Total income (dollars)	Family size	Age FM-1
Total food	4,100[a]	4,110	--	4	45
	1,073[a]	1,114	17,200	3	52
Food at home	1,069[a]	1,114	17,200	3	52
	1,019[a]	1,062	13,160	6	30
Cereal and bakery products	119	293	20,250	4	48
	115	426	7,560	5	31
Meat, fish, and poultry	1,032[a]	1,114	17,200	3	52
	1,012[a]	1,062	13,160	6	30
Dairy products	175[a]	517	3,200	5	35
	170[a]	460	8,040	8	44
Fruit	285	372	624	1	80
	213	274	11,172	1	82
Vegetables	76	878	150,000	4	31
	59	228	13,260	3	77
Sugar and sweets	67	216	3,072	1	71
	63[a]	124	14,500	4	45
Fats and oils	56	202	15,589	3	62
	44	200	4,750	1	68
Nonalcoholic beverages	117	391	11,060	3	26
	91	245	--	4	37
Other expenditures	232	621	27,500	3	75
	97	687	29,788	1	66
Food away from home	4,000[a]	4,110	--	4	45
	743	963	16,967	3	46
Meals	4,000	4,110	--	4	45
	699	755	--	1	40
Snacks	89	301	78,880	5	35
	55[a]	337	9,000	2	42
Beverages	356	963	16,967	3	46
	89	210	8,445	1	37
Alcoholic beverages	435	767	30,225	5	31
	282[a]	418	13,239	1	24
Tobacco and smoking supplies	123	331	7,678	8	48
	113	212	8,728	3	23
Personal care	731[a]	1,171	--	3	56
	208	350	2,160	4	30
Nonprescription medicines	519	556	1,885	1	78
	260	427	53,400	5	38
Housekeeping supplies	267	467	21,110	5	30
	227	687	29,788	1	66
Utilities and fuels	989[a]	1,142	11,060	2	80
	783[a]	882	--	2	61
Automobile fuel and lubricants	1,246	1,495	10,336	2	25
	577[a]	671	15,900	5	40
Other expenditures	822[a]	980	1,500	4	27
	640[a]	938	108,000	1	38
Total expenditures	4,110[a]	4,110	--	4	45
	1,496	1,496	10,336	2	45

[a]One-week expenditure adjusted to a two-week basis.

71

An evaluation of the 1972-73 BLS CEDS and 1972-73 BLS CEIS revealed 39 households in the CEDS and 38 in the CEIS whose total reported expenditures were more than 5 standard deviations from the mean. In this respect, the data meet normal expectations.[5] However, even those relatively few outliers may signal data problems for the researcher. Furthermore, examining particular categories of expenditures produces additional misgivings. The means, standard deviations, largest values, and number of CUs reporting expenditure greater than 5 standard deviations from the mean for the 9 major expenditure categories and for the most important food subgroups in the 1972-73 BLS CEDs are tabulated in Table 3.9. It is obvious that extreme values are easily encountered for almost any expenditure. The largest reported values in a category are generally 25 to 87 standard deviations above the mean expenditure of all CUs reporting nonzero values. Further examination of the CEDS food expenditures data reveal that the largest total values are usually due to one large subitem rather than larger across-the-board expenditures. Similar outliers are present in the 1972-73 BLS CEIS major expenditure categories, Table 3.10.

Further examples are provided in Tables 3.11 through 3.14. Tables 3.11 and 3.12 provide examples of two largest outliers

TABLE 3.12.   Details on two largest outliers for each major expenditure category, 1972-73 BLS CEIS

Expenditure category	Amount (dollars)	Total expenditure (dollars)	Total income (dollars)	Family size	Age FM-1
Food	17,449	23,172	13,139	6	47
	17,416	22,834	13,766	8	44
Alcohol	6,000	19,783	13,836	1	32
	3,760	11,857	17,692	2	50
Tobacco	2,028	23,089	--	2	54
	1,820	9,157	22,057	4	49
Housing	46,433	48,283	13,295	3	28
	36,682	46,470	8,269	3	39
House furnishings and equipment	12,907	72,259	--	4	48
	11,617	32,626	17,800	3	39
Clothing and material	11,463	24,177	15,006	1	37
	11,270	52,452	111,950	2	61
Transportation	39,690	72,259	--	4	48
	28,323	38,996	31,479	2	32
Medical care	31,239	99,716	131,321	5	53
	12,166	15,740	6,861	2	63
Personal care	2,356	7,110	7,853	2	58
	2,056	30,146	62,933	3	44
Recreation	21,596	24,889	5,340	1	72
	10,057	30,587	20,400	1	56
Reading	1,581	31,323	71,513	4	50
	1,559	23,794	14,262	5	44
Education	11,469	62,406	115,676	6	51
	10,155	99,716	131,321	5	53
Other expenditures	26,417	41,049	20,281	7	46
	5,236	16,470	23,529	2	61
Total expenditure	99,716	99,716	131,321	5	53
	72,259	72,259	--	4	48

TABLE 3.13. Descriptive statistics on specific expenditure categories as a percentage of total expenditures, 1972–73 BLS CEDS

Expenditure category	Mean proportion[a]	Standard deviation	Largest value	No. CUs > 5 S.D.[b]
Total food	57.2	18.6	100	0
Food at home	43.8	10.5	100	0
Cereal and bakery products	5.7	4.8	100	88
Meat, fish, and poultry	17.4	11.6	100	43
Dairy products	6.6	5.7	100	78
Fruit	3.6	3.7	100	78
Vegetables	3.9	3.2	67	78
Sugar and sweets	2.0	2.4	60	78
Fats and oils	1.8	1.9	59	71
Nonalcoholic beverages	3.8	3.4	100	68
Miscellaneous	4.1	4.0	100	85
Food away from home	17.5	15.4	100	--
Meals	16.2	14.9	100	--
Snacks	4.0	4.5	100	104
Beverages	1.8	3.2	86	52
Alcoholic beverages	7.8	8.8	100	52
Tobacco and smoking supplies	7.8	8.3	100	74
Personal care	6.5	7.4	100	88
Nonprescription medicines	4.3	8.2	100	87
Housekeeping supplies	5.1	5.0	100	98
Utilities and fuels	22.2	15.7	100	--
Automobile fuel and lubricants	15.2	11.5	100	74
Other expenditures	5.6	8.1	100	112

[a]Mean of those CUs reporting nonzero expenditures.
[b]Standard deviation.

for each major expenditure category from the 1972–73 BLS CEDS and the 1972–73 BLS CEIS, respectively.

Mean proportions of each expenditure category to total expenditure for those CUs reporting nonzero expenditures from the 1972–73 BLS CEDS and CEIS are presented in Tables 3.13 and 3.14, respectively. Standard deviation, largest value, and number of CUs reporting expenditures greater than 5 standard deviations of mean for each expenditure category are also shown in Table 3.13 and 3.14.

TABLE 3.14. Descriptive statistics on specific expenditures categories as a percentage of total expenditures, 1972–73 BLS CEIS

Expenditure category	Mean proportion[a]	Standard deviation	Largest value	No. CUs > 5 S.D.[b]
Total food	23.7	11.9	100	11
Alcohol	1.0	2.2	47	152
Tobacco	1.9	2.8	56	85
Housing	29.2	14.3	99	--
House furnishings and equipment	2.9	5.5	63	72
Clothing and material	7.1	5.1	83	45
Transportation	19.2	14.7	100	1
Medical care	6.6	6.6	100	90
Personal care	1.3	1.6	56	59
Recreation	3.4	3.8	87	70
Reading	0.6	0.9	19	106
Education	0.9	3.2	85	194
Other expenditures	0.9	2.6	64	174

[a]Mean of those CUs reporting nonzero expenditures.
[b]Standard deviation.

In conclusion, there are many potential model/data conflicts in the BLS consumer expenditure data bases. The researcher must give at least as much attention, thought, and diligence to the data base he or she intends to use as to the model he or she is trying to parameterize. It is not possible to say how many times hypotheses have been accepted or rejected because of the influence of an outlier on the estimates or because the researcher ignored the implicit assumption in the data being forced into a model.

Model-Data Interactions and Result Problems

King (1979) has described the consequences of ill-chosen models on the resulting model parameters. He notes Brandow's (1969) pleas for building research findings into a coherent whole. Brandow likened the accumulation of knowledge to trimming a Christmas tree. The individual research results are the ornaments that one attaches to the tree of hypothesized relationships. King emphasized the need for a coherent and relevant framework that will be useful for policy analysis. He further argued that the researcher has spent too much time testing uninteresting null hypotheses which he calls "village idiot" hypotheses, rather than establishing more relevant ones, and that the mathematical models chosen by the researcher are unrealistic because they conflict with the observations. He concluded that researchers have inhibited rather than expanded our knowledge of consumer demand. This section expands upon King's argument. Researchers interested in modeling consumer expenditure patterns have compounded their problems by failing to recognize a number of model/behavioral problems that may contribute to lack of success in this area.

Cross-sectional surveys have been available since the 1955 BLS Consumer Expenditures Survey and the 1955 USDA Household Food Consumption Survey. The vast detail in the cross-sectional surveys makes them a very fertile source of hypotheses and tests of models of consumer behavior. Yet researchers thus far have not utilized the potential of these surveys. Published research does not even begin to sum to any meaningful whole. In reviewing the published literature, it seems that the present framework (theoretical model) can most charitably be called an interesting Bonsai rather than a full-fledged tree. At present, no models adequately describe real-world consumer expenditure behavior. The literature describes many interesting attempts at parameterization of demand models pruned by repeated applications of the ceteris paribus assumptions.

Many may argue that we have good estimates of complete systems. George and King's (1971) work on food and Houthakker and Taylor's (1970) work on the complete array of household expenditures are examples. More recent ones are by Lee and Phillips (1971), Peterson (1972), and Chang (1979). However, results differ from study to study even when using the same data base. Some of these studies may suffer from the "omitted relevant variable" problem. The resulting estimates from such limited-information models produce biased elasticity estimates with the degree and direction of the bias depending upon the variables and observations omitted.

The order of magnitude of the difference encountered is seen in a comparison of two studies, both utilizing the 1965-66 USDA Household Food Consumption Survey (1965-66 USDA HFCS). George and King, controlling for family size, obtain an estimate of the income elasticity of food expenditures of 0.309. (See George and King 1971, App. Table A-2.) In contrast, Salathe and Buse, who included education and race of the household head, the employment status of the female head, and family size and composition, estimate the expenditure elasticity for food to be 0.228. (See Salathe and Buse 1979, Table 11.) That is a 25 percent difference between the two estimates. Furthermore, Salathe and Buse show large variations in income elasticities for differing levels of education, employment status, and race. A few examples are illustrated in Table 3.15.

TABLE 3.15. Estimated income elasticities based on level of education, race, and employment status of female head of household, 1965 USDA Household Food Consumption Survey

| Characteristic | Elasticities | | |
	Total food	Beef and pork	Dairy products
Educational level			
Less than 8 years	.258	.300	.100
8 to 11 years	.208	.291	.163
12 to 15 years	.195	.272	.147
16 or more years	.242	.304	.122
Employment status			
Employed	.212	.288	.126
Not employed	.229	.299	.151
Race			
White	.217	.281	.137
Black	.262	.332	.189
Other	.134	.356	.162
Sample average	.226	.297	.146

Source: Salathe and Buse (1979, 18).

Two studies using the BLS 1960-61 Consumer Expenditure Survey also display differences (Table 3.16). Chang (1979, Table 5.16) estimates the income elasticities for total food expenditures to be 0.56; Houthakker and Taylor (1970, Table 6.5) calculate it as 0.51. Though that estimate is close, others differ substantially. Chang's estimate of the food-away-from-home expenditure elasticity is twice that of Houthakker and Taylor (1.24 v.s. 0.58). Differences exist as well for the family size elasticities possibly because Chang controls many more socioeconomic characteristics than Houthakker and Taylor,

or because of Houthakker and Taylor's use of the means of grouped data to parameterize their model. In any case, the differences are sufficient to give anyone intending to use the results
for policy purposes a great deal of reason for confusion and
hesitancy.

TABLE 3.16. Comparison of estimated income and family size
elasticities, selected expenditures, 1960–61 BLS CEDS

Expenditures	Income		Family size	
	Chang[a]	H and T[b]	Chang[a]	H and T[b]
Total food	0.564	0.513	.487	.332
Food away from home	1.240	0.578	−.383	.293
Clothing	1.066	1.021	.107	.338
Personal care	0.786	0.832	.016	.079
Transportation	1.581	1.470	−.155	−.190
Recreation	1.159	1.120	.084	.092
Household operation	0.979	0.918	−.008	−.086
Medical care	0.796	0.697	.186	−.187

[a]Chang (1979, 188–89).
[b]Houthakker and Taylor (1970, 260–63).

Geographic differences. Chang has tested and rejected the hypothesis that the Engel functions are equal for urban households
in each of 9 U.S. cities. The rejection implies that the elasticities in Table 3.16 vary significantly from city to city.
Some of his results are compared to those of Houthakker and
Taylor's estimates in Table 3.17. More recently, Blaylock and
Smallwood (1980) examined the relationship of food elasticities
to income distribution. They used the 1972–73 BLS Consumer
Expenditure Diary Survey data and concluded: "the (food)
elasticities were found to vary substantially across income
classes, urbanization, and food groups" (p. 10).

At present, there is no basis for developing an overall
complete picture from the piecemeal results available and no way
to check consistency across different research reports. Nor is
there a way to determine the reliability of the estimates, such
as their agreement with total food disappearance or with reesti-
mates of models across several data sets. The unexplained differences across various published research reports make it
impossible to choose among the estimates. Thus, there is an
urgent need for researchers to recognize the importance of a
much wider range of socioeconomic variables on their estimates
in order to control more carefully these differences and one
hopes to produce more consistent and useful results.

Relevance of variables. Researchers using cross-sectional data
sets have made little serious effort to control the many relevant variables that account for at least part of the household-
to-household variation in expenditures. Thus, the omitted
relevant variable problem impugns the validity of current
parameter estimates. Even a casual analysis of the data sets
negates the researcher's assumption that included and excluded

TABLE 3.17. Comparison of estimated income and family size elasticities, selected U.S.
    cities, 1961-62 BLS CEDS

					Chang[a]						Houthakker
	Pooled	Boston	New York	Wash. D.C.	Cleve- land	Chi- cago	At- lanta	Dal- las	Los Angeles	Hono- lulu	and Taylor[b]
Income elasticities											
Total food	0.56	0.47	0.54	0.46	0.40	0.46	0.76	0.43	0.54	0.51	0.51
Personal care	0.79	0.61	1.03	0.86	0.72	0.85	0.96	0.88	0.73	1.01	0.83
Transpor- tation	1.58	1.56	1.37	1.75	1.77	1.82	1.65	1.82	1.93	1.61	1.47
Recreation	1.16	0.98	0.96	1.02	1.29	1.16	1.11	0.99	1.55	1.30	1.12
Household operation	0.98	0.95	1.19	1.01	0.81	0.92	1.23	0.93	0.89	0.45	0.92
Medical care	0.80	0.62	0.95	0.78	0.62	0.88	0.79	1.10	0.96	0.66	0.70
Clothing	1.07	1.11	1.26	1.07	1.00	1.12	0.99	1.13	0.92	0.88	1.02
Family size elasticities											
Total food	0.49	0.55	0.56	0.51	0.63	0.54	0.29	0.45	0.54	0.59	0.33
Personal care	0.02	0.16	0.11	0.10	0.14	0.06	0.10	0.20	0.08	-0.06	0.08
Transpor- tation	-0.15	-0.18	-0.15	-0.17	-0.25	-0.22	-0.10	-0.38	-0.20	-0.14	-0.19
Recreation	0.03	0.09	0.16	0.11	-0.11	-0.02	0.18	0.04	-0.07	0.05	0.09
Household operation	-0.07	-0.05	-0.08	-0.07	-0.03	-0.10	-0.12	-0.12	0.02	0.03	-0.09
Medical care	0.19	0.37	0.13	0.12	0.16	0.14	0.05	0.38	0.34	0.28	-0.19
Clothing	0.11	0.26	0.08	0.14	0.18	0.04	0.12	0.15	0.08	-0.11	0.34

[a]Chang (1979).
[b]Houthakker and Taylor (1970).

variables are uncorrelated. The assumption may make the
analysis simpler, but it also makes the results much less useful.
    Even in cases where researchers exhibit concern for the
mitigating effects of race, age, education, occupation, region,
and so forth, the variables are usually included as simple
dummies affecting the intercept. Yet, even casual observation
reveals that households with different socioeconomic characteristics exhibit different marginal propensities. And few researchers explicitly recognize the likely nonlinear nature of the
income elasticity which that approach implies. In other words,
the very common assumption is made that the income elasticities
are the same for all values of the exogenous sociodemographic
variables. Estimates, descriptions, and models which do not
recognize these interdependencies and interactions are far less
satisfactory than those which do.

Data utilization. One reason for the "slowness" of researchers
to challenge the mining of the available cross-sectional surveys
is that the requisite data handling techniques have been slow to
develop. Only within the past decade has sufficient computational power (via computers) been available to handle large data
bases. Only within the past five years has the computer software been generally available to researchers to manipulate the
large cross-sectional data bases. Unfortunately, the technologies permit the researcher to remove herself or himself from
direct contact with the data rather than use its ability to help

him or her become much more intimately familiar with what it
reveals about consumer behavior.

In summary, we are now in a state where theoretical models
definitely are lagging behind our available data. Models are
highly oversimplified explanations of how the household operates. In contrast, the data in the various consumer expenditure
surveys reflect the full spectrum of consumer behavior. Trying
to fit present models to present-day cross-sectional survey data
is akin to putting round pegs into square holes. They fit if
the square hole is large enough, but the fit leaves much space
unfilled.

The foregoing does not imply that theory has nothing to
contribute. On the contrary, the researcher should use theory
where it can help but not limit conceptualizations and imagina-
tion by its shortcomings. One must work on improved explanation
and description of _actual_ household expenditures. This, in
turn, can be the motivation for future theoretical models. This
approach is tedious and requires more data. It demands diligence and keen powers of observation plus the patience to test
and retest the results on new data sets. The approach may be
more fruitful in that successful descriptive models may motivate
the theoretician to develop better theoretical models of consumer behavior. The effort can only be successful in terms of
obtaining research funds and in the resulting increase in knowledge if it builds upon past and current work in a coherent
way. Old models must be retested with new data. Results must
be compared across data sets, models, and methods. It also
means abandoning much of our excess theoretical baggage and
striking out in new directions. Theoretical models are no more
than hypothesized relationships that require testing and retest-
ing before their validity can be accepted. They can only be
tested properly if the assumptions of the models and those implicit in the data used to test them coincide. We need good
descriptive analyses of household expenditures to use as input
into our theories. Accomplishing this will require particular
attention in research to the following areas.

Functional Forms

Observations suggest to us that a household's demand may be
affected by a wide variety of socioeconomic and demographic
characteristics. There is also research evidence to support
that proposition. The problem is how best to incorporate the
characteristics into one's descriptive model. The more recent
advances in theoretical models offer the researcher some potentially fruitful avenues of research for incorporating the sociodemographic into theoretical demand systems. In a recent paper,
Howe (1977) extends the Linear Expenditure System (LES) and the
Extended Linear Expenditure System (ELES) so that the subsistence quantities are functions of the household's sociodemographic characteristics. In principle, his approach can be

extended to include changing marginal budget shares across household characteristics.

A more general approach to this problem, scaling, was first proposed by Barten (1964) and tested by Muellbauer (1977) as a systematic method of incorporating household compositional effects into demand analysis. Recently, Pollack and Wales (1978b) proposed translating as a general method for incorporating demographic variables into systems of demand equations. In both translating and scaling, the original demand system is replaced by a new set which contains parameters suitable for introducing sets of conditioning or interacting variables. The new parameters are then specified to be functions of the socio-demographic variables and incorporated into the original demand system before it is estimated. Translating is illustrated by incorporating family size into a quadratic expenditure system (QES) and the LES. Pollack and Wales also demonstrate that all of the parameters of both the QES and the LES can be identified with two periods of a cross-sectional data sets. Since the Bureau of Labor Statistics will be releasing the continuous quarterly panel expenditure data for economists' use this method merits further exploration. In another report, Pollack and Wales (1978a) compared demographic translation and scaling in incorporating family size and age composition into three complete demand systems, the QES, the Basic Translog (BTL), and the Generalized Translog (GTL). Since they used British household expenditures data, generalizations are not possible.

Incorporating scaling and translation functions into demand models may be a fruitful approach, provided they do not conceal other model/behavioral inconsistencies are not concealed. Obviously, a careful examination of the implicit assumptions in the manner of King is required before one can reach a final conclusion as to the contribution of these techniques to an improved theoretical framework.

Improved functional forms. The appropriate functional form is very important for reliable estimates of the structure one is trying to describe and understand. Again, economic theory does not provide much guidance to the researcher as to the shape of the demand structure. The choice is left up to the investigator and he or she usually takes one of two options: arbitrarily choosing one functional form, or trying alternative functions and, based on certain pre-established criteria, choosing the "best one."

The most frequently chosen forms are linear, semi, and double log. Although there are reasonable arguments for the different functional forms, each imposes important a priori restrictions on the results.[6] There is a great need for a systematic approach for finding the most useful functional form across household expenditures. Again, cross-sectional surveys are a good source of data. They usually have sufficient observations to subdivide the sample to permit better statistical testing.

Prais and Houthakker (1955) considered a number of functional forms and chose among them on an ad hoc basis. They show that the expenditure elasticities for a particular product can vary by 50 percent at the mean under different functional forms. More recently, Salathe (1978) examined alternative functional forms and concluded that "the choice of the functional form has a dramatic impact on estimated income and household size elasticities." Kulshreshtha (1978) examines several functional forms in studying the demand for red meats in Canada. He concludes: The functional form for any commodity should not be left to the discretion of the investigator. Furthermore, the investigator should not limit the choice between linear versus logarithmic (or some other nonlinear) function form, but should open his search to any functional form (p. 48).

Recently, using a transformation procedure suggested by Box and Cox (1964), Benus et al. (1976) showed that a very general Engel function can be estimated. A disadvantage of using different functional forms for different expenditures is that the prediction of the individual expenditure categories will not necessarily sum to total expenditures. Leser (1963) states that a desirable property of Engel curves is that they meet the "adding-up" criteria. This means that the sum of all k individual expenditures must equal total expenditures. In the past, the imposition of this constraint has forced researchers to use the same functional form for all expenditure categories. Nicholson (1957) shows that when nonlinear functions are fitted to data that satisfy the Engel aggregation criteria, the resulting estimates will also satisfy the criteria within the observed data range. Thus, the Box-Cox procedure warrants further investigation as a method of introducing more flexibility into the functional form describing household expenditures.

An alternative approach that also seems to have potential is the Multinomial Logit Model of expenditure proportions. It allows Engel functions for different goods to exhibit distinctly different form and still meet the Cournot and Engel aggregation criteria. In addition, it permits incorporating sociodemographic characteristics and imposing other standard theoretical constraints.

Interaction variables. There is an additional troublesome aspect of functional form. It concerns the manner in which the researcher incorporates explanatory variables into the model. The most common assumption is that they enter linearly and additively. Again, this probably conflicts with actual behavior. There are likely to be interactions among the independent variables. An interaction means that the effect of one predictor on the dependent variable depends on the level of other predictors. Alternative names for interactions include conditioning effects, contingency effects, moderator effects, and specification effects. For example, it seems reasonable to expect that a change in income will be distributed differently for a southern rural family than for a northern urban one, for a college educated household head than for one with a high school education. This is supported by Salathe and Buse's work with the 1965 USDA Household Food Consumption Survey and the 1960-61 BLS Consumer Expenditure Surveys. In both data sets, income interacts with a wide range of variables including race, education, marital status, occupation, and family size. Thus, if we are to truly understand the structure of U.S. household expenditures, interactions must be seriously considered as part of our models.

Zero expenditures. There are many arguments for disaggregating expenditures into more homogeneous groups. Earlier discussion presented ample evidence that the further one disaggregates, the more likely one is to encounter zero expenditures. This leads to a problem that has not received much attention but which plagues researchers--low data density (i.e., zero expenditures for substantial subsets of observations). If the researcher is working with a linear model, zero expenditures may be scattered all along the values of the independent variable and the resulting parameter estimates would likely be biased. The degree of bias is difficult to ascertain without knowing the extent and location of the zero values in the data swarm.

Another problem with zero expenditures is that many of the most useful hypothesized functional forms are inconsistent with zero valued observations. If the researcher ignores this gross rejection by the data of his or her hypothesized functional form and proceeds to parameterize the model, there are four choices, (1) adding a positive constant to all sample observations, (2) replacing zero observations with a small positive constant, (3) excluding all observations that contain zero values, (4) replacing the logarithm of a zero-valued observation with a small positive value. The effect of each of these alternatives on the resulting OLS estimates are not well known. Johnson and Rausser (1971) show that using alternatives 1 to 3 on the explanatory variables leads to biased and inefficient estimates. They suggest an interactive estimation procedure. However, in estimating Engel functions it is zero expenditures on the dependent variable that are most worrisome. There is no theoretical work to guide the researcher.

Commodity groupings. A final problem is that of grouping commodities. With the wealth of information in the recent consumer surveys there is a temptation to estimate very disaggregative expenditure relationships. The 1972-73 Consumer Expenditure Survey contains data on 2,300 different expenditure items and the 1972-73 Diary, more than 180 foods and 80 nonfood expenditures. The theoretical literature develops the concept of separability, but it is of little or no practical help to the researcher faced with aggregating 2,300 different expenditures into theoretically acceptable but practicable groups. De Janvry (1966) and Young (1977) tried factor and cluster analysis and Bieri tried Cannonical Correlation to group commodities in cross-sectional surveys. Their results are difficult to interpret. On the other hand, arbitrary groupings based upon previous work or casual empiricism is not very satisfactory, either. It is also an area that needs further work.

PUBLIC USE DATA BASES

The public cost of planning and implementing surveys of tens of thousands of households is enormous. The resulting data are also practically priceless. USDA and BLS are to be commended for recognizing the value of the data for purposes other than their primary justification to the respective agencies. If researchers are to have access to future data sets and make the best use of them, open lines of communication between researchers and the agencies responsible for the data generation are essential in order that each other's needs are understood. Toward that goal the following points deserve emphasis.

Nondisclosure

The application of nondisclosure regulations needs review. Some means of maintaining the narrow line between infinite disclosure of detail and the gross aggregation used to comply with nondisclosure rules is essential. Both producers and users accept the great cost of such surveys, their usefulness for improving the understanding of household behavior, and the comparatively low cost of making the data available to researchers. A cost-benefit analysis would clearly favor allowing bona fide researchers more detail in certain critical variables. Identity of households could still be restricted. Researchers can assist in locating possible data inconsistencies that might arise and suggest means of these corrections and future avoidance.

The original 1972-73 BLS public use tapes (CEDS and CEIS) did not disclose either income information on CUs reporting a gross income of less than $2,000 or greater than $35,000 or details of family members in households with seven or more members (Table 3.18). If the 1965 survey nondisclosure decision had prevailed, researchers would have had to work with about 3,000 less observations in the Diary Survey and 2,500 less in the Interview Survey. The efficiency of our parameter estimates would have decreased disproportionately because 13 percent of the total numbers of households in the CED and CES represents the tails of the distributions (i.e., the largest and smallest incomes and the largest households).

Other information gaps require filling. The 1972-73 BLS CEDS and CEIS public use tapes contained information on more than 2,800 expenditure items but insufficient detail on the geography of the CU, the head's education and race, and any

TABLE 3.18. Comparison of number of comsumer units with regard to income
disclosure, and income and expenditures reporting status, 1972–73 BLS
CEDS and CEIS

Level of reporting on original tapes	Diary survey		Interview survey	
	Number	Percent	Number	Percent
Completed and income disclosed[a]	17,383	75.0	16,312	81.7
Completed but income not disclosed[b]	3,014	13.0	2,574	12.9
Incomplete reporting	2,789	12.0	1,089	5.4
Total consumer units	23,186	100.0	19,975	100.0

[a]Includes households with annual income between $2,000 and $34,999 and
less than 7 members which were in the original BLS CEDS and BLS CEIS data tapes.
[b]Includes households with annual income less than $2,000, or households
with annual income of $35,000 or more and households with 7 or more members
which were omitted from the original BLS CEDS and BLS CEIS data tapes.

product or service price information. There is strong evidence
to reject the hypothesis that prices are constant across households within a region in the 1972–73 BLS CEDS. Chang (1979)
reports similar differences among 9 major cities in the 1961–62
BLS CEDS. Hassan and Johnson (1977) also report similar findings with Canadian data.

Improved Public Use Tapes

The researcher must always exercise great care, diligence,
and skepticism in using any data set. However, experience
indicates that much data checking duplication could be eliminated by increasing the resources devoted to planning these activities and involving the advice and counsel of interested researchers in producing the machine readable data sets. Frequently, small changes or additions obtained at insignificant
marginal cost can increase the research usefulness of the data
base immensely. Addition of the education of the spouse and
information on food stamp purchases in the first year of the
1972–73 BLS Diary are two examples. Information on characteristics of the male head of the household in the 1965 USDA Household Food Consumption Survey is a third, and elimination of
alphabetic codes from numeric fields is a fourth. It appears
that so far there is little or no researcher input into the
design of the survey document developed by BLS and USDA.

Data reorganization and cleaning are major areas of
concern. Experience with the public use tapes over the past
seven or eight years indicates that agencies producing the tapes
(either BLS or USDA) employ specialists with little research
experience of the type for which the PU tapes are most widely
used. They produce tapes requiring a great deal of manipulation
and massaging before they can be used efficiently. The careful
researcher has to (1) carefully check the internal consistency
of the data items within each observation, and (2) modify, re-
define, and reorganize the information to reduce computing costs
and data churning. Even before statistical computing systems
such as SPSS, BMDP, or SAS can be used to analyze the PU data

tape, its format must be changed. The detail PU tapes for the BLS Diary Survey and the Interview Survey are in a format that produces a variable number of records per CU observation. Thus, one household with 3 expenditures has 3 expenditure records while another reporting 50 different expenditures has 50 records. In the Diary Survey a given CU has a potential of more than 250 different expenditures and over 2,800 records in the Interview Survey. This type of data file cannot be processed with any of the most widely used statistical computing systems noted above. For each researcher to repeat this data consolidation task leads to expensive duplication of effort and computational time.

In contrast, there are other areas where the level of aggregation has probably been carried too far. Marital status, race, education, and housing tenure are examples. Reporting the marital status of the CU head as married or not married is not satisfactory. The simple married-not married code is insufficient because different types of households are likely to exhibit quite different expenditure patterns, depending upon whether the head is divorced, separated, or widowed. Similarly, reporting race of household head as black or not black is another loss of detail whose inclusion would have cost little. Finally, reporting education in six categories rather than actual years violates the statistical principle of not throwing away information. Using categorical data where continuous data exist is always less satisfactory. Reasons for such aggregative categories is not obvious since more detailed categories would not have strained the coding capacity of a single column code.

Clearly, what is needed in the future is a structure for eliciting researcher requirements and for evaluating trade-offs.

SUMMARY

The time has arrived for systematically developing models which have superior power to explain household to household variation in demand or expenditures. The recent upsurge in the technologies with which data can be economically generated, the wealth of data coming on line, and the low cost methods of handling large complex data sets give rise to optimism, but also to potential dangers. Complex microeconomic models that first explain and then simulate household demand behavior with some degree of detail are within the realm of feasibility if researchers avoid the data traps and make efficient use of the information in these data bases. The release of the 1977-78 USDA Nationwide Food Consumption Survey and the BLS continuous expenditure panel are further encouraging signs.

NOTES

1. Author is Barry W. Bobst.
2. Author is Rueben C. Buse.
3. See Buse for details on the data tape and its construction.

4. Tchebysheff's theorem proves that at least 1 - 1/(k**2) of the total observations should lie within 3 standard deviations of their mean. Thus, no more than 1/9 or approximately 2,500 observations in the BLS Diary should be more than 3 standard deviations from the mean and about 930 should be more than 5 standard deviations from the mean. Similarly, in the Survey no more than 2,220 should lie beyond 3 standard deviations from the mean and no more than 800 observations should be 5 or more standard deviations from the mean.

5. These expenditures were adjusted to a two-week basis on Tables 3.9 and 3.10 to provide comparability with other reported expenditures.

6. See Brown and Deaton (1972), Hassan and Johnson (1977), or Salathe (1978) for detailed discussion of the strengths and weaknesses of alternative functional forms of Engel functions.

REFERENCES

AAEA Committee on Economic Statistics. 1972. Our obsolete data system: New directions and opportunities. Am. J. Agric. Econ. 54:867-75.

Aigner, D. J. 1974. MSE dominance of least squares with errors-of-observations. J. Econometrics 2:365-72.

Barten, A. P. 1964. Family composition, prices and expenditure patterns. In Econometric analysis for national economic planning: 16th Colston Symposium, ed. P. E. Hart, G. Mills, and J. K. Whitaker. London: Butterworth.

Benus, J., J. Kmenta, and H. Shapiro. 1976. The dynamics of household budget allocation to food expenditure. Rev. Econ. and Stat. 57:129-38.

Bieri, J. H. 1969. The quadratic utility function and measurement of demand parameters. Ph.D. diss., Univ. of California, Berkeley.

Blaylock, J. R., and D. M. Smallwood. 1980. The relationship of food expenditures to income distribution. USDA/ESCS/NED pap., Washington, DC.

Bonnen, J. T. 1975. Improving information on agriculture and rural life. Am. J. Agric. Econ. 57:753-63.

Box, G. E. P., and D. R. Cox. 1964. An analysis of transformations. J. R. Stat. Soc. Series B, 26:211-43.

Brandow, G. E. 1969. Economic research needs for public policy in food and fiber. Proceedings: A seminar on better economic research on the U.S. food and fiber industry. USDA, ERS, 8-14.

Brown, J. A. C., and A. S. Deaton. 1972. Surveys in applied economics: Models of consumer behavior. Econ. J. 82:1145-236.

Buse, R. C. 1979. Data problems in the BLS/CES PU-2 diary tape: The Wisconsin 1972-73 CES diary tape. Agric. Econ. Rep. No. 164, Dep. of Agric. Econ., Univ. of Wisconsin-Madison.

Chang, O. H. 1979. Impact of permanent income, prices and sociodemographic characteristics on household expenditure patterns in the United States, 1960–61. Ph.D. diss., Univ. of Wisconsin–Madison.

DeJanvry, A. 1966. Measurement of demand parameters under separability. Ph.D. diss., Univ. of California, Berkeley.

Garber, S., and S. Klepper. 1980. Extending the classical normal errors–in–variables model. Econometrica 48:1591–546.

George, P. S., and G. A. King. 1971. Consumer demand for food commodities in the U.S. with projections for 1980. Giannini Found. Monogr. No. 26, Univ. of California, Berkeley.

Hassan, Z. A., and S. R. Johnson. 1977. Urban food consumption patterns in Canada. Agric. Can., (Ottawa), Econ. Br. Publ. 77/1,

Hayenga, M. L., and D. Hacklander. 1970. Monthly supply–demand relationships for fed cattle and hogs. Am. J. Agric. Econ. 52:535–44.

Hooper, J. W., and H. Theil. 1958. The extension of Wald's method of fitting straight lines to multiple regression. Rev. Int. Stat. Inst. 26:37–48.

Houthakker, H. S., and L. D. Taylor. 1970. Consumer demand in the United States 1929–1970. 2d ed. Cambridge, MA: Harvard Univ. Press.

Howe, H. 1977. Cross–section application of linear expenditure systems: Responses to sociodemographic effects. Am. J. Agric. Econ. 59:141–48.

Johnson, S. R., and G. C. Rausser. 1971. Effects of misspecifications of log–linear functions when sample values are zero or negative. Am. J. Agric. Econ. 53:120–24.

King, R. A. 1979. Choices and consequences. Am. J. Agric. Econ. 61:839–48.

Kulshreshtha, S. N. 1978. Demand for red meats in Canada: A search for functional form. Tech. Bull. No. 78–3, Dep. of Agric. Econ., Univ. of Saskatchewan, Saskatoon, Canada.

Lee, F. Y., and K. E. Phillips. 1971. Differences in consumption patterns of farm and nonfarm households in the United States. Am. J. Agric. Econ. 53:573–82.

Lee, J. E., Jr. 1972. Discussion: Our obsolete data systems. Am. J. Agric. Econ. 54:875–77.

Leser, C. E. V. 1963. Forms of Engel functions. Econometrica 31:694–703.

Maddala, G. S. 1979. Disequilibrium economics. Pap. presented at Am. Agric. Econ. Assoc. Meeting, Pullman, Washington, July 30.

Malinvaud, E. 1970. Statistical methods of econometrics. 2d rev. ed. New York: American Elsevier.

Morgenstern, O. 1963. On the accuracy of economic data. 2d ed. Princeton: Princeton Univ. Press.

Muellbauer, J. 1977. Testing the Barten model of household consumption effects and the cost of children. Econ. J. 87:460–87.

Musgrove, P. 1978. Consumer behavior in Latin America: Income
    and spending of families in ten Andean cities. Washington:
    Brookings Inst.
Nicholson, J. L. 1957. The general form of the adding-up cri-
    terion. J. R. Stat. Soc. 120:84-85.
Pearl, R. B. 1977. Data systems on food demand and consumption:
    Properties, uses, future prospects. In Food Demand and
    Consumption Behavior, ed. R. Raunikar, Athens: S-119
    Southern Regional Research Committee and the Farm Foundation.
Peterson, H. P. 1972. Estimating the influence of household
    composition   on   household   food   expenditures   by
    adult-equivalent scales for households in the United States
    in 1955 and 1965. Ph.D. diss., Univ. of Wisconsin-Madison.
Pollack, R. A., and T. J. Wales. 1978a. Comparison of the qua-
    drastic expenditure system and translog demand system with
    alternative specifications of demographic effects. Univ. of
    British Columbia, Discuss. Pap. No. 78-21.
_____. 1978b. Estimation of complete demand systems from house-
    hold budget data: The linear and quadratic expenditure
    systems. Am. Econ. Rev. 68:348-59.
Prais, S. J., and H. S. Houthakker. 1955. The analysis of
    family budgets. Cambridge: Cambridge Univ. Press.
Salathe, L. E. 1978. A comparison of alternative functional
    forms for estimating household Engel curves. Pap. at
    meetings of Am. Agric. Econ. Assoc., Virginia Polytechnic
    Inst. and State Univ.
_____. 1979. Household expenditure pattern in the U.S.   USDA/
    ESCS Tech. Bull. No. 1603.
_____, and R. C. Buse. 1979. Household food consumption
    patterns in the United States. USDA/ESCS Tech. Bull. No.
    1587.
Theil, H. 1971. Principles of econometrics. New York: Wiley.
Tyrrell, T. J. 1979. An application of the multinomial logit
    model to predicting patterns of food and other household
    expenditures in the Northeastern United States. Ph.D.
    diss., Cornell Univ.
U.S. Department of Agriculture. 1972. Major statistical series
    of the U.S. Department of Agriculture. Agricultural
    Handbook No. 365, Vol. 5. Washington, DC: U.S. Government
    Printing Office.
_____. 1975. Agricultural statistics 1975. Washington, DC:
    U.S. Government Printing Office.
_____. 1975. Agricultural statistics 1979. Washington, DC:
    U.S. Government Printing Office.
U.S. Department of Commerce. 1977. 1974 Census of agriculture.
    Vol. 1, part 51. Washington, DC: U.S. Government Printing
    Office.
Wold, H. O. A., and L. Jureen. 1953. Demand analysis: A study
    in econometrics. New York: John Wiley.
Young, T. 1977. An approach to commodity grouping in demand
    analysis. J. Agric. Econ. 28:141-51.

PART II

Complete Demand Systems

Comparison of Estimates from Three Linear Expenditure Systems

IN CHAPTER 1, the theoretical framework of consumer demand was developed. By explaining the concepts of neoclassical demand theory and derivation of important theoretical properties of the system of demand functions, Chapter 1 provided the foundation for the empirical application and estimation of consumer demand to be discussed in this chapter. Following the theoretical development of Chapter 1, the purpose of this chapter is to focus on functional form specifications and empirical results obtained from three alternative systems of demand functions. Parameter estimation of the systems of demand functions are reported and compared, and their relative strengths and weaknesses are evaluated.

Specifically, discussions presented in this chapter are organized as follows: (1) linear expenditure systems, (2) elasticity formulas, (3) specification of demand categories, (4) data characteristics and sources, (5) estimation methods, (6) comparison of results among estimated systems, and (7) conclusions.

LINEAR EXPENDITURE SYSTEMS

Among a wide variety of complete demand systems, the linear expenditure systems have been very popular and often used in empirical studies.[1] Three commonly employed versions of the linear expenditure system are presented in this section: (1) Stone's linear expenditure system, (2) Leser's approximation, and (3) Powell's approximation.

In Chapter 1, it was shown that for a given set of commodity prices, a set of demand equations can be derived by maximizing a utility function subject to the income or budget constraint. If

---

Authors of this chapter are John A. Craven and Richard C. Haidacher.

the resulting demand functions are assumed to be linear in all prices and income, and expressed in the expenditure form, then the set of demand functions can be written as

$$p_i q_i = c_i + \sum_{j=1}^{n} a_{ij} p_j + \beta_i y \qquad (i = 1, \ldots, n) \qquad (4.1)$$

Here P_i and q_i are the price of and quantity demanded for the ith commodity, respectively, $p_i q_i$ is the expenditure on the ith commodity, c_i is the ith intercept, the a_{ij} are price parameters, β_i is the marginal budget share for the ith commodity, and Y represents the consumer's income. A system of functions such as (4.1) for the n commodities is called a linear expenditure system (LES). Homogeneity of degree zero in prices and total expenditure can be preserved in any LES by setting all intercept terms to zero.

Stone's Linear Expenditure System

In 1954, Stone published a landmark article that contained parameter estimates for a LES. Stone's LES has the property that the general restrictions of classical demand theory are satisfied exactly over each set of data coordinates, the only LES that has this property. Since Stone's LES is the only LES to globally satisfy the general restrictions of classical demand theory, it has become known in demand literature as The Linear Expenditure System. The term "Stone's LES" is used to avoid confusion with various approximations to the classical LES.

This version of the LES is obtained from the constrained maximization of the Klein-Rubin utility function, so named because it is implicit in an article by Klein and Rubin (1947-48). It is also called the Stone-Geary utility function. This utility function is of the form

$$U = \sum_{i=1}^{n} \beta_i \ln(q_i - \Upsilon_i) \qquad (4.2)$$

where the β_i and the Υ_i are parameters.

Maximization of (4.2) subject to the traditional budget constraint yields the demand equations of the system

$$q_i = \Upsilon_i + (\beta_i/p_i)(y - \sum_{j=1}^{n} p_j \Upsilon_j) \qquad (4.3)$$

Thus, the expenditure functions[2] which constitute Stone's LES are

$$p_i q_i = p_i \Upsilon_i + \beta_i (y - \sum_{j=1}^{n} p_j \Upsilon_j) \qquad (4.4)$$

This system has been and continues to be very popular among empirical analysts. One of the reasons for its popularity stems from the intuitive appeal of the interpretation that can be given to the expenditure functions. The interpretation, attributable to Samuelson (1948), is that the Y_i (if positive) are those quantities which the consumer perceives to be "minimum requirements" or in some sense "necessary." Given this interpretation, the expenditure on the ith commodity consists of the expenditure on the minimum required quantity of the ith commodity plus the proportion of the budget which is left over after the expenditure on all minimum requirements is accounted for. This proportion, β_i, is the marginal budget share, and the dollar amount to be allocated is called "supernumerary" income.

An advantage of deriving demand equations from a specific utility function is that the general restrictions of classical demand theory are automatically satisfied at each set of data coordinates. A disadvantage of this approach is that the theoretical implications inherent in the choice of a particular form may prove to be unduly restrictive. The choice of the Klein-Rubin utility function has a number of implications. (A discussion of the properties of the elasticity estimates is deferred until later.) These include

1. The Y_i must be smaller than the corresponding q_i at each point in time in order for logarithms of (4.2) to be defined.

2. Since the marginal utility for each commodity $[\beta_i/(q_i - Y_i)]$ is positive, the respective marginal budget share (β_i) must also be positive. Hence, inferior goods are precluded.

3. The marginal utility of each good is independent of the level of consumption of all other goods, a general property of all "strongly separable" utility functions.

4. The expression for the Slutsky substitution terms is positive in all cases $[k_{ij} = (\partial Y_i/\partial p_j)^* = \beta_i(q_j - Y_j)/p_i, \ i \neq j]$. Consequently, complementary goods are precluded.

Leser's Approximation

Leser's approximation represents the parameterization of (4.1) in terms of the Hicks-Allen partial elasticities of substitution (Powell 1974, 12-14, 63-66; Hicks and Allen 1934). These elasticities are defined by the relation $\sigma_{ij} = \dot{e}_{ij}/w_j$, where σ_{ij} is the partial elasticity of substitution of the ith commodity with respect to the jth commodity, $\dot{e}_{ij}$ is the income compensated elasticity of the ith commodity with respect to the price of the jth commodity, and w_j is the budget share of the jth commodity ($p_j q_j/y = w_j$).

Partial elasticities of substitution are related to the

Slutsky substitution terms by the expression $\sigma_{ij} = k_{ij}y/(q_iq_j)$.
Given the symmetry of the Slutsky substitution matrix, the symmetry of the partial elasticity of substitution matrix is apparent from this formulation.

Leser's approximation involves rewriting (4.1) in terms of the income-compensated cross-price elasticities, substituting these relationships into the expression for the partial elasticities of substitution, and finally solving for the price parameters (a_{ij}) of (4.1) in terms of the partial elasticities of substitution. Consequently, the solutions for the a_{ij} are

$$a_{ij} = \bar{w}_i\bar{q}_j\bar{\sigma}_{ij} - \beta_i\bar{q}_j + \delta_{ij}\bar{q}_i \tag{4.5}$$

where $\delta_{ij} = 1$ when $i = j$, 0 otherwise. The bars indicate the coordinate set for which the partial elasticities of substitution are evaluated (e.g., the sample means). Upon substitution of (4.5) into (4.1), Leser obtains

$$r_i = \sum_{j \neq i} \bar{\sigma}_{ij}\epsilon_{ij} + \beta_i v \tag{4.6}$$

as an approximation to the classical LES, where

$$r_i = p_iq_i - p_i\bar{q}_i \tag{4.7}$$

$$\epsilon_{ij} = w_i\bar{p}_j\bar{q}_j(p_j/\bar{p}_j - p_i/\bar{p}_i) \tag{4.8}$$

$$v = y - \sum_{j=1}^{n} p_j\bar{q}_j \tag{4.9}$$

This system satisfies the adding-up and homogeneity restrictions of classical demand theory over all sets of data coordinates. It also satisfies Slutsky symmetry at a <u>selected</u> set of data coordinates. Leser's final assumption is that <u>all</u> partial elasticities of substitution are equal, allowing the system to be simplified further

$$r_i = \bar{\sigma}\phi + \beta_i v \tag{4.10}$$

where $\phi = \bar{w}_i \sum_{j=1}^{n} (p_j\bar{q}_j - p_i\bar{q}_i)$.

Powell's Approximation

Powell's approximation to the classical LES utilizes a property of strongly separable utility functions in order to reduce the number of price parameters to be estimated. A general characteristic of strongly separable utility functions is that all cross-substitution effects of price change are related to their income derivatives by a proportionality variable (ψ) at each set of data coordinates

$$k_{ij} = - \psi(\partial Y_i/\partial y)(\partial Y_j/\partial y) \qquad (i \neq j) \tag{4.11}$$

where ψ is a function of all prices and total expenditure.[3] Powell assumes that the value of ψ evaluated at the means of income and prices is an adequate approximation to ψ over the entire range of price and income variation ($\bar{\psi} = \psi$). Proceeding in a manner similar to Leser and solving for the a_{ij} of equation (4.1), Powell's solutions are

$$a_{ii} = (1 - \beta_i)(\bar{\psi}\beta_i/\bar{p}_i + \bar{q}_i)$$
$$a_{ij} = - \beta_i(\bar{\psi}\beta_j/\bar{p}_j + \bar{q}_j) \tag{4.12}$$

Substituting these relationships into the homogeneous version of (4.1), Powell obtains the transformed system[4]

$$r_i = \bar{\psi}\Xi + \beta_i v \tag{4.13}$$

where r_i and v are identical to Leser's definition and

$$\Xi = \beta_i \sum_{j=1}^{n} \beta_j (p_i/\bar{p}_i - p_j/\bar{p}_j) \tag{4.14}$$

Powell's transformation has the properties of adding-up and homogeneity of degree zero over all sets of data coordinates. In addition, the matrix of Slutsky substitution terms is not only symmetric at the sample means of the data, but also any point where the price ratios p_i/p_i are equal to the ratio of mean prices $\bar{p}_i/\bar{p}_j$.

The gains from the imposition of the classical demand theory constraints and the additional restrictive assumptions of the various systems are as follows. In an n equation system, equation (4.1) implies that there are $n(n + 2)$ parameters to be estimated. In Stone's LES, this number is reduced to $2n - 1$. Both Leser's and Powell's approximations further reduce this number to n.

ELASTICITY FORMULAS
Ordinary own- and cross-price elasticity formulas for the

three systems are presented in Table 4.1 (the usual <u>ceteris paribus</u> caveat is assumed throughout the discussion on elasticities). None of the systems are characterized by constant elasticities. It is likely that there is a different elasticity for each set of data coordinates. The elasticity formulas for the Leser and Powell systems are simplified if the elasticities are evaluated for the set of data coordinates which correspond to the set selected for the application of the theoretical restrictions ($p_i = \bar{p}_i$, $q_i = \bar{q}_i$, etc.). Subsequent discussion of the Leser and Powell system elasticity estimates is confined to those obtained for the mean set of data coordinates. Statements regarding the Leser and Powell elasticities are not, in general, applicable to all sets of prices and quantities.

TABLE 4.1. Price elasticity formulas

	Ordinary own-price	Ordinary cross-price
Stone	$e_{ii} = -1 + (1 - \beta_i)\gamma_i/q_i$	$e_{ij} = -\beta_i(p_j\gamma_j/p_iq_i)$
	$e_{ii} = -1 + (1 - \beta_i)\bar{q}_i/q_i - (1 - \bar{w}_i)\bar{\sigma}\bar{q}_i/q_i$	$e_{ij} = (\bar{w}_i\bar{\sigma} - \beta_i)p_j\bar{q}_j/p_iq_i$
Leser	$e_{ii} = -\beta_i - (1 - \bar{w}_i)\bar{\sigma}^a$	$e_{ij} = (\bar{w}_i\bar{\sigma} - \beta_i)\bar{p}_j\bar{q}_j/\bar{p}_i\bar{q}_i^a$
	$e_{ii} = -1 + \beta_i\bar{\psi}(\beta_i - 1)/\bar{p}_iq_i + (1 - \beta_i)\bar{q}_i/q_i$	$e_{ij} = (\beta_i\beta_j\bar{\psi}/p_iq_i)p_j/\bar{p}_j - \beta_i(p_j\bar{q}_j/p_iq_i)$
Powell	$e_{ii} = \beta_i\bar{\psi}(\beta_i - 1)/\bar{p}_i\bar{q}_i^a - \beta_i$	$e_{ij} = (\beta_i\beta_j\bar{\psi}/\bar{p}_i\bar{q}_i) - \beta_i(\bar{p}_j\bar{q}_j/\bar{p}_i\bar{q}_i^a)$
	Compensated own-price	Compensated cross-price
All	$\dot{e}_{ii} = e_{ii} + \beta_i$	$\dot{e}_{ij} = e_{ij} + \beta_i(p_jq_j/p_iq_i)$

aEvaluated at the mean value of the variables.

Some general statements can be made about the algebraic signs of the price elasticities. For Stone's LES, all ordinary price elasticities are negative. The ordinary own-price elasticity for the Leser system is negative and the ordinary cross-price elasticity is negative provided $\beta_i > w_i\bar{\sigma}$. For the Powell system, the ordinary own-price elasticity is negative and the ordinary cross-price elasticity is negative provided $p_j\bar{q}_j > \beta_i\bar{\psi}$.

Income-compensated own-price elasticities are negative and income-compensated cross-price elasticities are positive for each of the three systems. Total expenditure elasticities are the same for each system, namely $E_i = \beta_i/w_i$.

SPECIFICATION OF DEMAND CATEGORIES

The restrictions inherent in these systems have implications for the specification of the commodities to be analyzed. The assumption of strongly separable utility in the Stone and Powell systems implies that the marginal utilities of each commodity should be independent of the levels of consumption of all other commodities. Also, the positive signs on the income-compensated cross-price derivatives indicate that all commodities should be substitutes in order to be analyzed within the framework of these systems.

The following broad "commodities" or consumption categories are seemingly consistent with the implicit restrictions of the models:

1. Food for off-premise consumption (food home)
2. Food for on-premise consumption (food away)
3. Alcohol and tobacco (alc.-tob.)
4. Clothing (cloth.)
5. Housing (house)
6. Utilities (util.)
7. Transportation (trans.)
8. Medical care expenses (medical)
9. Durable goods (dur. gds.)
10. "Other" nondurable goods (n. dur. gds.)
11. "Other" services (services)

Several complete demand system studies use demand categories which correspond to the major groups published in U.S. Department of Commerce's Survey of Current Business (1979).[5] Major differences between the above grouping and others are the separation of the food component into at-home and away-from-home categories, the exclusion of alcohol from the food categories, the delineation of medical expenses into a separate category, and the inclusion of automobile expenditures in the durable goods category. Expenditures on the individual items included in the eleven categories do not sum to total personal consumption expenditure as reported in Survey of Current Business. A twelfth category, "other" miscellaneous, consisting largely of items which either are not a direct component of consumer expenditure or do not pass through the marketplace, is excluded from the analysis. A detailed listing of the individual items in each consumption category is contained in Table A4.1 at the end of this chapter.

DATA CHARACTERISTICS AND SOURCES

The data used in the estimation procedure were annual observations for the 1955-78 time period. The primary data source is the personal consumption expenditure series reported in the Survey of Current Business by the United States Department of Commerce.[6] Commodity expenditure is represented by per capita personal consumption expenditure measured in current dollars (current $PCE). Quantities are represented by per capita personal consumption expenditures measured in 1972 dollars (constant

$PCE). Constant $PCE can be used to represent quantity since
the variation in this series is due to the variation in quantities only. Resident population data for the United States were
used to place each expenditure series on a per capita basis.
These data were obtained from U.S. Bureau of Census, Population
Estimates and Projections (1979). For prices, implicit price
deflators (IPD), obtained from the identity, current $PCE/con-
stant $PCE = IPD, were used. The data are exhibited in Tables
A4.2, A4.3 and A4.4 at the end of the chapter.

ESTIMATION METHODS
    Leser and Powell both include linear trend terms in applica-
tions of their models. Thus equations (4.10) and (4.13) are
modified by the inclusion of the term $d_i t$; d_i is the trend parameter for the ith commodity and t is a variable representing
time. Time is measured in a negative to positive direction with
the value zero assigned to the middle of the data set. This
procedure preserves homogeneity of degree zero in prices and
total expenditure only at the midpoint of the data set.
    The estimation procedures developed by Leser (1960), and
Powell (1966) are utilized to estimate their respective models.[7]
Rewriting (4.10) with the trend term and applying Leser's set of
restrictions to the data means yields

$$r_i = \bar{\sigma}\phi + \beta_i v + d_i t \qquad\qquad (4.15)$$

If the average value for the elasticity of substitution were
known, the term $\bar{\sigma}\phi$ could be moved to the left-hand side of (4.15)
and a system which had identical explanatory variables would result. Leser proceeds to estimate $\bar{\sigma}$ by minimizing the grand total
sum of squares over all equations. He then makes the transformation suggested above and estimates the resulting system by applying ordinary least squares to each equation.[8] The adding-up condition $\Sigma\beta_i = 1$ and $\Sigma d_i = 0$ is automatically satisfied by the
ordinary least squares method.
    The estimation of Powell's system is more complex. By
rewriting (4.13) with the trend term

$$r_i = \bar{\psi}\Xi + \beta_i v + d_i t \qquad\qquad (4.16)$$

and recalling the definition of Ξ, it is evident that Powell's
system involves nonlinearities among $\bar{\psi}$, β_i, and the β_j. Powell
(1966) proceeds in a manner similar to Leser. First, he notes
that if $\bar{\psi}$ and Ξ were known, he could use Leser's method by moving

$\bar{\psi}\Xi$ to the left-hand side of (4.16). Powell then employs Leser's
estimator for the β_i to obtain Ξ, obtains a value for $\bar{\psi}$ by using
Leser's method to estimate $\bar{\sigma}$, moves $\bar{\psi}\Xi$ to the left-hand side of
(4.16), and estimates the β_i and d_i of (4.16) by ordinary least
squares. This set of β_i is then used to obtain a new Ξ vector
and the process is repeated until the parameter estimates stabi-
lize around constant values.

   Stone's LES also requires a nonlinear estimation technique
due to the presence of the product terms involving β_i and the γ_j
in equation (4.4). To estimate Stone's LES, a nonlinear,
maximum likelihood estimation program written by Snella (1979)
was used. Snella's algorithm uses a gradient technique to
search for the maximum of the likelihood function.[9]

   Maximum likelihood estimation of complete demand systems is
complicated by the fact that the adding-up condition implies a
singular disturbance covariance matrix for each time period.
Snella's algorithm is capable of estimating the parameters of
any system in which the disturbances are independently and
identically normally distributed with zero expectation and a
constant covariance matrix with rank n − 1.

   Regarding the statistical properties of the estimators,
those of the Stone system will be consistent and asymptotically
efficient. The estimators in the Leser system are best linear
unbiased if $\bar{\sigma}$ is unbiased. The properties of the Powell estimator cannot be established since values for the β_i appear on
both sides of the equation to be estimated. In subsequent discussion of the summary statistics for the Leser and Powell models
it should be kept in mind that these are only approximations
based on ordinary least squares formulas.

COMPARISON OF RESULTS AMONG ESTIMATED SYSTEMS
   The explanatory ability of the estimated equations appears
to be excellent for each of the three systems. The lowest
proportion of "explained" variation in the expenditure on any
commodity (as measured by R^2 values) was 0.98. Thus, the R^2
values did not provide any basis for judging the superiority of
one system over the others.

   Estimates of marginal budget shares obtained from the three
models are presented in Table 4.2. Average budget shares (w_i)
are presented for comparison purposes. Each of the estimates is
large compared to its standard error; and the values for each
commodity, with a few notable exceptions, appear reasonably
close to each other from model to model. The marginal budget
share for food at home ranges from 0.0349 for the Powell system
to 0.0621 for the Leser system with the Stone system estimate

lying between the two. On the basis of the associated standard
errors, one would have more confidence in the Stone or Leser
estimate; but the aforementioned precaution about the properties
of these standard errors prevents rejecting any system estimate
on this basis. Perhaps the widest discrepancy in the model
estimates is for the nondurable goods category. Here, the Stone
system estimate is more than twice the value for the other two
systems. Again, all system estimates for this demand category
are large relative to their standard errors.

TABLE 4.2. Marginal budget shares, parameter estimates from three
        linear expenditure systems

Category	Leser	Powell	Stone	w_i
Food home	.0621	.0349	.0472	.1623
	(.0071)	(.0126)	(.0048)	
Food away	.0472	.0443	.0302	.0427
	(.0035)	(.0035)	(.0018)	
Alc.-tob.	.0093	.0166	.0158	.0505
	(.0015)	(.0021)	(.0007)	
Cloth.	.0725	.0837	.0723	.0988
	(.0039)	(.0037)	(.0016)	
House	.1693	.1760	.1948	.1473
	(.0078)	(.0085)	(.0031)	
Util.	.0550	.0509	.0427	.0390
	(.0034)	(.0028)	(.0010)	
Trans.	.0955	.0988	.0907	.0811
	(.0048)	(.0056)	(.0018)	
Medical	.1498	.1545	.1427	.0752
	(.0078)	(.0081)	(.0031)	
Dur. gds.	.2063	.2083	.1897	.1323
	(.0161)	(.0157)	(.0051)	
N.dur. gds.	.0289	.0283	.0644	.0556
	(.0056)	(.0066)	(.0035)	
Services	.1040	.1099	.1093	.1152
	(.0044)	(.0047)	(.0018)	

Note: Top number is parameter estimate; numbers in parentheses
are associated standard errors.

    The remainder of the directly estimated parameter estimates
are presented in Table 4.3. Referring first to the estimates of
the "minimum requirement" parameters of the Stone system (Y_i),
our first observation is that all are quite large when compared
to their asymptotic standard errors. The largest estimate is
for the food at home category followed by services and housing.
Comparing the estimates with the "quantity" data series, the Y_i
for food at home is larger than the 1955 observation, and the Y_i
estimate for nondurable goods is larger than the 1955 through
1958 observations on their respective demand categories. Thus,
the Klein-Rubin utility function cannot be strictly interpreted
given our results. In experimenting with data sets of different
lengths, we have noticed that it is not uncommon to obtain some
estimates for the "minimum requirements" which are larger than
the first few observations in the data set. Usually, when the

time frame is shortened this discrepancy disappears. This result suggests that the hypothesis of a fixed value for the minimum requirement parameter may be too restrictive, especially for time-series which exhibit strong trends.

TABLE 4.3. Additional parameter estimates, three linear expenditure systems

| Category | Trend coefficients | | Y_i |
	Leser	Powell	Stone
Food home	-0.1979	0.8138	410.01
	(0.5482)	(0.9700)	(6.02)
Food away	-0.9891	-1.0179	106.23
	(0.2720)	(0.2684)	(2.14)
Alc.-tob.	0.3589	-0.0683	137.87
	(0.1133)	(0.1611)	(0.94)
Cloth.	-0.6662	-1.0179	226.81
	(0.2994)	(0.2811)	(2.91)
House	1.7069	1.7450	271.33
	(0.5958)	(0.6496)	(6.83)
Util.	-0.8851	-0.6214	72.34
	(0.2587)	(0.2146)	(1.31)
Trans.	-0.2945	-0.1466	162.40
	(0.3658)	(0.3674)	(3.56)
Medical	-0.8600	-0.8772	120.85
	(0.5976)	(0.6230)	(5.82)
Dur. gds.	-1.7534	-1.7399	220.88
	(1.2313)	(1.1958)	(9.67)
N.dur. gds.	2.8507	3.0389	108.51
	(0.4296)	(0.5049)	(5.30)
Services	0.7274	-0.1105	272.34
	(0.3341)	(0.3612)	(4.75)

$$\bar{\sigma} = 0.46624 \qquad Y = 802.4075$$

Note: Top number is parameter estimate; numbers in parentheses are associated standard errors.

The estimates for the trend parameters of the Leser and Powell systems can be directly compared. There are three cases in which the algebraic signs on the coefficients differ for the two systems. These instances occur for food at home, alcohol-tobacco, and services. In each of these cases, the trend coefficient for the Powell system is small relative to its standard error. The trend coefficient for the food at home equation in the Leser system is small when compared to its standard error but the coefficients for the alcohol-tobacco and services equations appear to be of acceptable magnitude when compared to their standard errors. In the food away, housing, medical, durable goods, and nondurable goods equations, the two systems yield trend coefficients which are similar in magnitude and are of the same algebraic sign. In each of these cases, the parameter estimates are large relative to their standard errors. This result gives some degree of confidence that reliable estimates were obtained for these parameters in both the Leser and Powell systems.

The estimate for the (constant) elasticity of substitution in the Leser system is 0.47. This result indicates a substant-

ial degree of curvature in the indifference curves between any
two commodities since a zero value would imply right-angled
indifference curves and a value of infinity would imply flat
indifference curves. The method by which $\bar{\sigma}$ and $\bar{\psi}$ were estimated
does not permit their standard errors to be calculated.

Estimates of total expenditure elasticities for the three
systems calculated at the sample means (1955-78) are presented
in Table 4.4.

TABLE 4.4. Total expenditure elasticities, Leser, Powell, and Stone
system estimates

Category	Leser	Powell	Stone
Food home	0.40	0.22	0.31
Food away	1.10	1.03	0.73
Alc.-tob.	0.20	0.35	0.34
Cloth.	0.76	0.88	0.78
House	1.14	1.18	1.38
Util.	1.39	1.29	1.13
Trans.	1.16	1.13	1.16
Medical	1.84	1.90	1.92
Dur. gds.	1.54	1.56	1.49
N.dur. gds.	0.52	0.50	1.20
Services	0.89	0.94	0.98

Observations similar to those made for the marginal budget
shares can be made for the total expenditure elasticity estimates due to the fact that the total expenditure elasticity
formulas for the three systems are the same. Since the average
budget shares are the same for the three systems, differences in
the total expenditure elasticities are due only to differences
in the system estimates for the marginal budget shares.

For each of the three systems, the estimate for the food at
home total expenditure elasticity is smaller in value than the
estimate for the food away from home total expenditure elasticity. However, the Stone system estimate yields the interpretation that food away from home is a necessity. This category
appears as a luxury in both the Leser and Powell systems.

Food at home is the most inelastic commodity with respect to
total expenditure delineated in the Powell and Stone systems,
while alcohol-tobacco is the most inelastic commodity delineated
in the Leser system. Medical expenses are the most elastic with
respect to total expenditure in each of the systems.

Leser's system own-price elasticities calculated at the
sample means are presented in Table 4.5. For this system, all
own-price elasticities are negative and lie in the range -0.6102
to -0.4534. Experimenting with various values for the constant
elasticity of substitution (selected a priori), it was found
that the Leser system own-price elasticity estimates are usually
quite close to the negative of the value selected for the constant elasticity of substitution. An examination of the formula
for own-price elasticities (evaluated at the sample mean) sheds

some light on this observation. If the value for $\bar{w}_i\bar{\sigma}$ does not differ substantially from β_i, the negative of the values for all own-price elasticities will lie very near the selected value of $\bar{\sigma}$. The assumption of a constant elasticity of substitution is probably too restrictive for the estimation of price elasticities of demand and, as a result, this system is rejected for estimation of own-price elasticities.

TABLE 4.5.   Uncompensated own-price elasticities, Leser's expenditure system

Category	Elasticity
Food home	-.4556
Food away	-.4935
Alc.-tob.	-.4534
Cloth.	-.4944
House	-.5663
Util.	-.5028
Trans.	-.5234
Medical	-.5781
Dur. gds.	-.6102
N.dur. gds.	-.4689
Services	-.5159

Uncompensated own- and cross-price elasticity estimates for the Stone and Powell systems calculated at the sample means are presented in Table 4.6. In this table, the items listed across the top of the page refer to prices and the items listed in the far left column refer to quantities. In general, the two systems yield elasticity estimates which are similar in magnitude. Exceptions to this statement occur for the food at home, food away from home, and nondurable goods categories.

In comparing the own-price elasticities one observes that for food at home, the Stone system estimate is about 27 percent larger than the Powell system estimate.[10] For food away from home, the Powell system estimate is about 57 percent larger than the Stone system estimate. Indeed, for the food at home and nondurable goods categories, all Stone system estimates are larger in absolute value than their Powell system counterparts. For both systems, the own-price elasticity estimate for food away from home is larger than for food at home.

The demand for medical services, durable goods, and housing are the most responsive to own-price changes. Food at home and alcohol-tobacco are the least responsive to own-price changes. Utilities and transportation, whose individual components consist largely of energy related commodities, have own-price elasticities which lie in the middle of the range.

With the exception of medical services and housing, the own-price elasticities agree with intuition. One might expect that the demand for medical services and housing would be more inelastic than certain other nonfood items in the consumer

TABLE 4.6. Uncompensated price elasticities, Stone and Powell expenditure systems (Powell data in parentheses)

Demand categories	Food home	Food away	Alc.-tob.	Cloth.	House	Util.	Trans.	Medical	Dur. gds.	N.dur. gds.	Services
Food home	-.1246 (-.0978)	-.0103 (-.0067)	-.0131 (-.0096)	-.0224 (-.0158)	-.0275 (-.0218)	-.0082 (-.0055)	-.0168 (-.0124)	-.0117 (-.0082)	-.0232 (-.0163)	-.0113 (-.0108)	-.0255 (-.0189)
Food away	-.1032 (-.1508)	-.2110 (-.3320)	-.0313 (-.0443)	-.0534 (-.0731)	-.0655 (-.1008)	-.0194 (-.0256)	-.0401 (-.0572)	-.0280 (-.0377)	-.0553 (-.0755)	-.0269 (-.0498)	-.0608 (-.0873)
Alc.-tob.	-.0475 (-.0508)	.0113 (-.0104)	.1015 (-.1164)	-.0246 (-.0247)	-.0302 (-.0340)	-.0089 (-.0086)	-.0185 (-.0193)	-.0129 (-.0127)	-.0255 (-.0254)	-.0124 (-.0168)	-.0280 (-.0294)
Cloth.	-.1108 (-.1284)	-.0264 (-.0264)	-.0336 (-.0377)	-.2652 (-.3186)	-.0703 (-.0858)	-.0209 (-.0218)	-.0430 (-.0487)	-.0300 (-.0321)	-.0594 (-.0642)	-.0289 (-.0424)	-.0652 (-.0743)
House	-.1951 (-.1726)	-.0465 (-.0355)	-.0591 (-.0507)	-.1009 (-.0837)	.4881 (-.4600)	-.0367 (-.0293)	-.0758 (-.0655)	-.0529 (-.0432)	-.1046 (-.0864)	-.0509 (-.0570)	-.1148 (-.1000)
Util.	-.1603 (-.1875)	-.0382 (-.0385)	-.0486 (-.0551)	-.0829 (-.0910)	-.1017 (-.1253)	-.3238 (-.4062)	-.0623 (-.0712)	-.0434 (-.0469)	-.0859 (-.0938)	-.0418 (-.0619)	-.0943 (-.1086)
Trans.	-.1642 (-.1641)	-.0391 (-.0337)	-.0498 (-.0482)	-.0849 (-.0796)	-.1042 (-.1097)	-.0309 (-.0279)	-.3605 (-.3900)	-.0445 (-.0410)	-.0880 (-.0821)	-.0428 (-.0542)	-.0967 (-.0950)
Medical	-.2723 (-.2766)	-.0648 (-.0568)	-.0825 (-.0812)	-.1408 (-.1342)	-.1729 (-.1848)	-.0513 (-.0470)	-.1058 (-.1050)	-.5569 (-.6215)	-.1459 (-.1385)	-.0710 (-.0914)	-.1603 (-.1602)
Dur. gds.	-.2108 (-.2273)	-.0502 (-.0467)	-.0639 (-.0668)	-.1090 (-.1103)	-.1338 (-.1519)	-.0397 (-.0386)	-.0819 (-.0863)	-.0571 (-.0568)	.5175 (-.5676)	-.0550 (-.0751)	-.1241 (-.1316)
N.dur. gds.	-.1699 (-.0731)	.0405 (-.0150)	-.0515 (-.0215)	-.0879 (-.0355)	-.1079 (-.0489)	-.0320 (-.0124)	-.0660 (-.0278)	-.0461 (-.0183)	-.0911 (-.0366)	-.3563 (-.1702)	-.1001 (-.0424)
Services	-.1393 (-.1376)	-.0332 (-.0283)	-.0422 (-.0404)	-.0720 (-.0668)	-.0884 (-.0920)	-.0262 (-.0234)	-.0541 (-.0522)	-.0377 (-.0344)	-.0747 (-.0689)	-.0363 (-.0455)	-.3318 (-.3545)

budget. The housing result may be due in part to data peculiar-
ities. Within the national accounts, the expenditure on housing
is calculated as an imputed rent for both owner-occupied and
tenant-occupied housing. Thus, for each year, homeowners are
assumed to spend an amount equal to the fair market rental value
on housing rather than an amount equal to their actual expenditures. A possible explanation for the medical services result
is that the demand for preventive and elective care has increased over the years and that this component of medical care is
more price responsive than the absolutely necessary component.
Of course, this is speculative.

Turning to a discussion of the cross-price elasticities, the
demand for food at home is more responsive to changes in prices
of housing and services than to changes in other nonfood prices.
It is least responsive to changes in prices of utilities and
food away from home. The demand for food away from home is most
responsive to changes in the price of food at home, followed by
changes in the prices of housing and services. The demand for
food away from home is substantially more responsive to a price
change in transportation than is the demand for food at home.
As was the case with the demand for food at home, a change in
the price of utilities has the least effect on the demand for
food away from home.

A general pattern emerges when the absolute values for the
cross-price elasticities are observed. A price change for any
of the items that are more important in terms of the total
budget (as measured by average budget shares) has a larger
percentage effect than does a price change for any item which is
of minor importance in the budget. An example illustrates this
point. Over the period of analysis, the average budget share
was 0.1623 for food at home and 0.0390 for utilities. The
cross-price elasticity for each demand category is consistently
larger for a change in the price of food at home than for a
change in the price of utilities. This is due to the dominance
of the income effect of a price change over the substitution
effect of a price change within the framework of these two
systems.

Plots of the observed "quantity" data series for food at
home and food away from home along with the estimates from the
three systems are presented in Figures 4.1 and 4.2. The estimated quantities were obtained by applying the system parameter
estimates to actual prices and total expenditure for each sample
year (see Table A4.5 for numbers upon which these figures are
based). A brief glance at Figure 4.1 seems to indicate that the
Leser system monitors the demand for food at home better than
the other systems. In fact, the Leser system approximates the
demand levels better than the other systems in 12 of the 24
sample years and identifies seven of nine turning points. The
Stone system approximates demand better in 7 of the 24 years and
identifies five turning points. For this purpose the Powell
system is clearly inferior to the other systems. Each of the

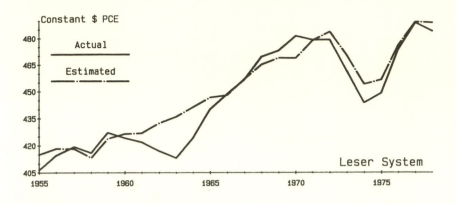

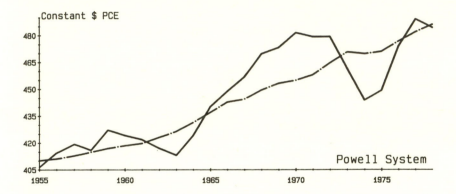

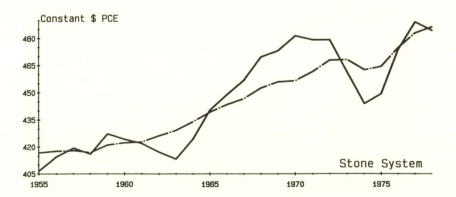

Fig. 4.1. Food at home, actual and estimated quantities, Leser, Powell, and Stone demand systems (1955-78).

106

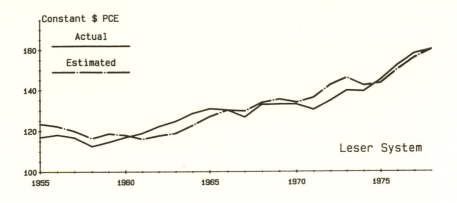

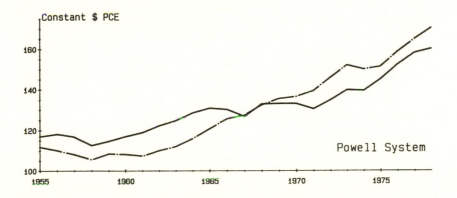

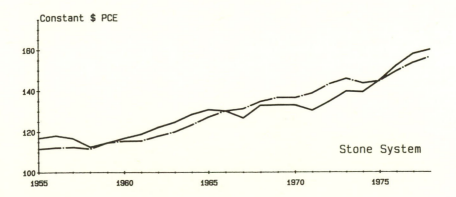

Fig. 4.2. Food away from home, actual and estimated quantities, Leser, Powell, and Tone demand systems (1955-78).

107

three systems missed the turning point in 1960--all systems
showed an increase in per capita consumption of food at home
through 1963 when it was, in fact, decreasing. Except for this
period, the Leser system performs quite well.

In Figure 4.2 a similar pattern emerges when the demand
system estimates for food away from home are compared to their
actual values. For this category, the Leser system yields
better estimates in 12 of the 24 sample years and identifies
seven of eight turning points. The Stone system performs better
in 8 years and identifies six turning points. The Powell system
is clearly outperformed by the other two systems but the gap
between the performance of the Leser and Stone system is
narrow--both appear to perform well.

CONCLUSIONS
The fact that the Leser system outperforms the other systems
in approximating food demand is a somewhat confusing result. On
the surface it would seem that the Leser system is the most
restrictive of the three systems in terms of the assumptions
made about its response parameters. Indeed, Leser's assumption
of a constant value for the (common) elasticity of substitution
and the importance of this parameter in the price elasticity
formulas is intuitively unappealing. There appears to be no
good reason to assume, a priori, that all system own-price
elasticities lie near to each other in magnitude.

Further insight into the relatively poor performance of the
Stone and Powell systems can be gained by examining the implicit
restrictions on price elasticities that are inherent in the
assumption of a strongly separable utility function. Deaton
(1974) has shown that a strongly separable utility implies that
all own-price elasticities are approximately related to their
total expenditure elasticities by a systemwide proportionality
constant (see Deaton and Muellbauer 1980, 138-39, for a discussion). This relation becomes nearly exact as the number of
commodities analyzed increases. With 11 consumption categories,
the results of this study lend empirical support to this hypothesized relationship. The absolute values of the Stone system
own-price elasticities are approximately one-third the size of
their corresponding expenditure elasticities. While the relationship is not as strong for the Powell system, ordering of the
commodities by the absolute value of own-price elasticities
corresponds to an ordering by the values of the total expenditure elasticities. Viewing consumer behavior in the context of
the optimizing model, there appears to be no a priori reason to
impose this proportionality restriction on the estimates of
demand parameters.

Consequently, the strengths of the three systems are also
their Achilles' heel in terms of the measurement of price elasticities of demand. The additional restrictions above those
inherent in the constrained maximization problem reduce the size
of the estimation problem considerably but also place undue

restriction on the price elasticities of demand. On a more positive note, it appears that the Leser and Stone systems merit further investigation as models for making demand projections.

APPENDIX

TABLE A4.1. Demand categories

Mnemonic	Description
Food home	Food purchased for off-premise consumption excluding alcohol.
Food away	Purchased meals and beverages excluding alcohol.
Alc.-tob.	Alcoholic beverages plus tobacco products.
Cloth.	Shoes and other footwear, shoe cleaning, repairing and accessories except footwear; cleaning, laundering, dyeing, pressing, alteration, storage, and repair of garments; jewelry and watches, other.
House	Owner or tenant occupied nonfarm dwellings.
Util.	Electricity, gas, fuel oil, and coal.
Trans.	Tires, tubes, accessories, and other parts; repair, greasing, washing, parking, storage, and rental; gasoline and oil; bridge, tunnel, ferry, and toll roads; insurance premiums less claims paid, purchased local transportation, purchased intercity transportation.
Medical	Drug preparation and sundries; physicians, dentists, other professional services; privately controlled hospitals and sanitariums; medical care, hospitalization insurance, income loss insurance, workmen's compensation insurance.
Dur. gds.	Furniture, including mattresses and bedsprings; kitchen and other household appliances; china, glassware, tableware, utensils, other durable house furnishings; books and maps; wheel goods, durable toys, sports equipment, boats and pleasure aircraft; radio and TV receivers, new autos, net purchases of used autos, other vehicles.
N.dur. gds.	Toilet articles and preparations; semidurable household furnishings; cleaning and polishing preparations, miscellaneous household supplies and paper products; stationery and writing supplies; magazines, newspapers, and sheet music; nondurable toys and sport supplies; flowers, seeds, and potted plants.
Services	Personal business expenditures; barbershops, beauty shops, and baths; water and other sanitary services; telephone and telegraph; domestic service, other household operation; radio and TV repair; admissions to spectator amusements; clubs and fraternal organizations; pari-mutuel net receipts, commercial participant amusements.
Other misc.	Private education and research; religious and welfare activities; net foreign travel; food furnished to employees; food produced and consumed on farms; clothing furnished by military; rental value of farm dwellings; other housing; ophthalmic products and orthopedic appliances.

TABLE A4.2. Personal consumption expenditures, aggregate demand categories, dollars per capita (1955–78)

Year	Food home	Food away	Alc.- tob.	Cloth.	House	Util.	Trans.	Medical	Dur. gds.	N.dur. gds.	Services
1955	271.6	62.9	85.8	169.2	192.2	56.2	111.0	76.4	210.2	72.1	147.3
1956	278.8	64.8	87.7	173.5	202.1	59.2	117.4	81.4	200.4	75.0	157.3
1957	291.5	66.5	89.7	172.1	212.7	61.7	123.6	87.8	203.7	78.2	163.8
1958	301.4	66.0	90.3	171.2	223.9	64.4	126.2	93.8	185.0	79.6	170.4
1959	305.1	69.2	95.6	177.8	235.9	65.9	134.1	100.6	210.9	84.2	181.2
1960	306.2	72.4	97.3	178.7	248.6	67.5	140.0	106.9	210.9	88.1	192.7
1961	307.5	75.2	98.4	179.5	260.1	68.1	140.7	112.4	198.7	91.7	200.7
1962	306.2	79.4	101.0	184.9	274.4	70.9	146.0	120.8	220.3	99.4	207.2
1963	307.3	82.9	103.8	187.5	286.3	73.6	149.1	128.3	240.4	104.8	218.1
1964	321.1	86.8	105.2	199.9	299.1	76.0	153.9	141.9	260.3	111.3	229.2
1965	341.1	90.4	109.5	207.9	315.0	79.0	163.4	149.3	288.2	118.9	243.2
1966	366.1	94.3	115.1	224.6	331.0	82.6	175.7	159.8	305.9	132.2	262.8
1967	371.0	96.5	119.4	232.7	348.9	86.1	186.3	171.6	311.5	137.5	284.4
1968	395.0	106.4	125.6	251.1	372.9	90.6	199.4	187.9	356.4	149.9	306.4
1969	417.2	113.1	130.4	267.3	400.6	96.1	216.2	215.0	376.8	180.9	329.4
1970	447.1	121.3	139.9	272.1	429.2	102.3	234.0	238.4	367.7	170.3	347.8
1971	453.6	125.0	145.0	288.5	466.4	109.5	253.5	259.3	419.1	179.8	368.7
1972	479.6	134.7	153.6	310.9	504.3	120.7	271.7	287.5	478.0	196.3	397.3
1973	532.7	151.4	161.1	341.8	548.0	134.8	295.8	318.6	526.2	219.0	429.6
1974	601.7	169.5	174.1	360.7	601.8	159.2	350.1	356.2	509.2	240.6	470.1
1975	655.8	191.4	186.0	384.3	655.3	185.4	373.6	410.3	549.7	252.3	516.8
1976	696.9	214.3	200.5	414.3	718.8	210.0	415.2	462.6	652.2	270.8	567.7
1977	746.2	237.6	209.7	446.4	803.0	237.4	461.3	529.5	737.1	292.3	615.9
1978	816.2	262.2	224.0	493.3	903.8	259.4	506.9	590.4	817.8	325.2	692.1

TABLE A4.3. Personal consumption expenditures, aggregate demand categories, constant (1972) dollars per capita (1955–78)

Year	Food home	Food away	Alc.- tob.	Cloth.	House	Util.	Trans.	Medical	Dur. gds.	N.dur. gds.	Services
1955	406.3	116.8	139.3	246.5	271.3	76.8	172.8	141.6	286.6	105.1	289.0
1956	414.4	118.0	140.9	247.5	280.3	79.5	179.3	147.8	265.5	107.3	292.8
1957	419.3	116.6	140.7	241.5	289.6	80.7	181.4	153.1	258.0	108.0	290.1
1958	415.7	112.4	140.4	239.2	299.5	83.3	180.0	158.3	233.6	107.2	291.4
1959	427.1	114.5	143.5	245.9	311.0	83.4	185.9	164.8	256.9	112.1	300.7
1960	424.1	116.8	142.1	243.4	322.9	84.1	189.5	170.0	255.8	115.4	306.0
1961	421.8	118.8	142.6	242.7	333.7	84.0	188.6	174.2	238.7	118.3	316.0
1962	416.9	122.1	144.6	248.8	347.8	86.7	193.9	183.7	260.4	127.4	316.8
1963	413.0	124.6	146.0	249.5	359.1	89.4	197.7	192.3	281.3	132.3	322.3
1964	424.2	128.4	145.8	263.1	371.4	93.0	203.9	208.4	301.6	138.2	331.7
1965	440.2	130.7	148.6	270.8	387.4	96.3	209.0	213.7	335.4	146.8	343.6
1966	448.7	130.0	151.8	285.4	401.4	99.7	218.5	219.2	355.3	161.6	356.2
1967	456.7	126.6	151.4	284.0	415.7	102.7	225.5	222.3	355.1	163.9	373.1
1968	469.9	132.8	151.6	291.1	433.9	106.5	235.7	231.5	391.6	172.0	378.9
1969	473.1	133.0	149.7	293.1	451.5	110.7	244.9	249.1	404.2	177.7	384.4
1970	481.3	132.9	151.1	286.8	464.8	113.4	250.8	263.3	384.6	181.0	385.8
1971	479.0	130.3	149.5	294.8	482.5	113.9	258.5	271.7	423.2	183.6	386.5
1972	479.1	134.7	153.6	310.9	504.3	120.7	271.7	287.5	478.0	196.3	397.3
1973	461.3	139.6	163.4	330.3	525.4	125.5	283.2	303.1	517.0	213.8	411.2
1974	443.6	139.2	162.9	325.9	529.3	118.2	282.8	310.8	468.7	209.8	415.1
1975	449.1	144.9	161.8	334.1	568.7	122.1	283.2	314.5	465.8	194.4	425.8
1976	473.7	152.1	166.5	348.4	591.7	126.2	291.3	330.6	522.6	197.4	446.5
1977	488.8	157.9	168.0	361.5	623.6	129.4	300.0	345.4	567.2	202.0	463.9
1978	484.0	159.9	169.0	387.2	656.6	131.6	314.3	354.4	595.5	213.5	487.0

TABLE A4.4. Implicit price deflators, aggregate demand categories (1955-78)

Year	Food home	Food away	Alc.-tob.	Cloth.	House	Util.	Trans.	Medical	Dur. gds.	N.dur. gds.	Services
1955	0.67	0.54	0.62	0.69	0.71	0.73	0.64	0.54	0.73	0.69	0.51
1956	0.67	0.55	0.62	0.70	0.72	0.75	0.66	0.55	0.76	0.70	0.54
1957	0.70	0.57	0.64	0.71	0.73	0.76	0.68	0.57	0.79	0.72	0.57
1958	0.73	0.59	0.64	0.72	0.75	0.77	0.70	0.59	0.79	0.74	0.59
1959	0.72	0.60	0.67	0.72	0.76	0.79	0.72	0.61	0.82	0.75	0.60
1960	0.72	0.62	0.68	0.73	0.77	0.80	0.74	0.63	0.83	0.76	0.63
1961	0.73	0.63	0.69	0.74	0.78	0.82	0.75	0.65	0.83	0.77	0.64
1962	0.73	0.65	0.70	0.74	0.79	0.82	0.75	0.66	0.85	0.78	0.65
1963	0.74	0.67	0.71	0.75	0.80	0.82	0.75	0.67	0.85	0.79	0.68
1964	0.76	0.68	0.72	0.76	0.81	0.82	0.76	0.68	0.86	0.81	0.69
1965	0.78	0.69	0.74	0.77	0.81	0.82	0.78	0.70	0.86	0.81	0.71
1966	0.82	0.73	0.76	0.79	0.82	0.83	0.80	0.73	0.86	0.82	0.74
1967	0.81	0.76	0.79	0.82	0.84	0.84	0.83	0.77	0.88	0.84	0.76
1968	0.84	0.80	0.83	0.86	0.86	0.86	0.85	0.81	0.91	0.87	0.81
1969	0.88	0.85	0.87	0.91	0.89	0.87	0.88	0.86	0.93	0.91	0.86
1970	0.93	0.91	0.93	0.95	0.92	0.90	0.93	0.91	0.96	0.94	0.90
1971	0.95	0.96	0.97	0.98	0.97	0.96	0.98	0.95	0.99	0.98	0.95
1972	1.00	1.00	1.00	1.00	1.00	1.00	1.00	1.00	1.00	1.00	1.00
1973	1.16	1.09	1.00	1.04	1.04	1.08	1.05	1.05	1.02	1.02	1.05
1974	1.36	1.22	1.07	1.11	1.10	1.35	1.24	1.15	1.09	1.15	1.13
1975	1.46	1.32	1.15	1.15	1.15	1.52	1.32	1.30	1.18	1.30	1.21
1976	1.47	1.41	1.20	1.19	1.22	1.66	1.43	1.40	1.25	1.37	1.27
1977	1.53	1.51	1.25	1.24	1.29	1.84	1.54	1.53	1.30	1.45	1.33
1978	1.69	1.64	1.33	1.27	1.38	1.97	1.61	1.67	1.37	1.52	1.42

TABLE A4.5. Food at home and food away from home, constant (1972) dollar
expenditures per capita, actual observations and systems estimates --
Leser, Powell, and Stone models (1955-78)

Year	Food at home				Food away from home			
	Actual	Leser	Powell	Stone	Actual	Leser	Powell	Stone
1955	406.3	414.9	410.1	416.6	116.8	123.3	111.6	111.5
1956	414.4	418.2	411.1	417.5	118.0	122.1	109.9	112.1
1957	419.3	418.1	412.8	417.8	116.6	119.8	107.9	112.3
1958	415.8	413.1	414.8	416.7	112.4	116.3	105.4	111.5
1959	427.1	423.8	416.9	421.0	114.5	118.6	108.3	114.6
1960	424.1	426.5	418.5	422.2	116.8	117.8	108.0	115.4
1961	421.8	426.9	419.7	422.6	118.8	115.9	107.2	115.5
1962	416.9	432.3	423.1	426.1	122.1	117.6	109.9	117.9
1963	413.0	436.0	426.4	429.0	124.6	118.8	111.9	119.9
1964	424.2	441.3	431.4	433.8	128.4	122.6	115.8	123.3
1965	440.2	446.7	437.0	439.1	130.7	126.9	120.6	127.1
1966	448.7	448.0	442.7	443.2	130.0	130.0	125.5	130.1
1967	456.7	457.0	444.3	446.5	126.6	129.6	127.4	131.2
1968	469.6	465.2	449.4	452.4	132.8	133.8	132.3	134.7
1969	473.1	468.9	453.1	455.8	133.0	135.4	135.4	136.6
1970	481.3	468.7	454.8	456.4	132.9	133.9	136.4	136.5
1971	479.0	478.7	457.8	461.4	130.3	136.2	139.2	138.7
1972	479.1	483.6	464.5	467.8	134.7	142.5	145.8	143.2
1973	461.3	470.1	470.5	468.1	139.6	146.0	151.8	145.8
1974	443.6	454.0	469.5	462.3	139.2	142.2	149.8	143.5
1975	449.1	456.6	470.8	464.2	144.9	143.3	151.1	144.6
1976	473.7	476.0	476.5	474.6	152.1	150.0	158.4	149.4
1977	488.8	489.3	481.7	482.8	157.9	155.8	164.7	153.5
1978	484.0	488.8	485.9	486.1	159.9	159.9	170.2	156.3

NOTES

1. Developed mainly from Powell (1974), Chapter 3.
2. These functions can also be obtained by substituting $a_{ii} = \Upsilon_i (1 - \beta_i)$, $a_{ij} = -\beta_i \Upsilon_j$, $c_i = 0$ in (4.1).
3. See Houthakker (1960) for a derivation of this result. A definition of ψ is $-\lambda/(\partial\lambda/\partial y)$ where λ is the marginal utility of money. The algebraic sign of ψ is positive.
4. In the homogeneous version $c_i = 0$ for all i.

5. The groupings correspond to those delineated in Mann (1980). Mann, however, uses twelve consumption categories in his system, although the parameter estimates for "other miscellaneous" are not reported.
6. Annual (1929 to present) and quarterly (1946 to present) personal consumption expenditure data are available on recorded tape from Computer System and Services Division, BEA, U.S. Department of Commerce, Washington, DC.
7. Appreciation is expressed to Rodney C. Kite for providing the computer program used to estimate the Leser and Powell systems.
8. Systems estimation of several equations which have identical explanatory variables will yield results identical with those obtained by applying ordinary least squares to each equation separately. See Theil (1971, 309).
9. A complete discussion of this technique is outside the scope of this chapter. For a complete discussion, the reader is referred to Snella (1979).
10. The practice of comparing elasticity estimates in terms of their absolute values is followed. Since all of the elasticity estimates have a negative algebraic sign, this practice should not cause confusion.

REFERENCES

Deaton, A. S. 1974. A reconsideration of the empirical implications of additive preferences. Econ. J. 84:338-48.
_____, and J. Muellbauer. 1980. Economics and consumer behavior. Cambridge: Cambridge Univ. Press.
Hicks, J. R., and R. G. D. Allen. 1934. A reconsideration of the theory of value, Parts I and II. Economica 1:52-76, 196-219.
Houthakker, H. S. 1960. Additive preferences. Econometrica 28:244-57.
Klein, L. R., and H. Rubin. 1947-48. A constant-utility index of the cost of living. Rev. Econ. Stud. 15:84-87.
Leser, C. E. V. 1960. Demand functions for nine commodity groups in Australia. Aust. J. Stat. 2:102-13.
Mann, J. S. 1980. An allocation model for consumer expenditures. Agric. Econ. Res. 32:12-24.

Powell, A. A. 1966. A complete system of consumer demand equations for the Australian economy fitted by a model of additive preferences. Econometrica 34:661–75.

_____. 1974. Empirical analytics of demand systems. Lexington, MA: Heath.

Samuelson, P. A. 1948. Some implications of linearity. Rev. Econ. Stud. 15:88–90.

Slutsky, E. [1915.] 1952. On the theory of the budget of the consumer. G. Econ. Trans. Olga Ragusa. In Readings in price theory, ed. G. J. Stigler and K. E. Boulding, Homewood, IL: Richard D. Irwin, Inc.

Snella, J. J. 1979. A fortran program for multivariate nonlinear regression--GCM. Univ. de Geneve, Cah. du Dep. d'Econ. Fac. des Sci. Econ. et Soc., Cah. 79.05.

Stone, R. 1954. Linear expenditure systems and demand analysis: An application to the British pattern of demand. Econ. J. 64:511–27.

Theil, H. 1971. Principles of econometrics. New York: Wiley.

U.S. Bureau of Census. 1979. Population estimates and projections, Ser. P–25, No. 802. Washington, DC: U.S. Government Printing Office.

U.S. Department of Commerce, Bureau of Economic Analysis. 1979. Survey of current business. Washington, DC: U.S. Government Printing Office.

CHAPTER 5

Persistence in Consumption Patterns: Alternative Approaches and an Application of the Linear Expenditure System

MUCH OF THE WORK in consumer demand analysis assumes that individuals adjust instantaneously to a new equilibrium when income and/or prices change. This approach is the static approach to demand analysis. In reality, the static theory is overly restrictive in that it ignores adjustments that occur through time due to habit formation, purchases of durable goods, and so forth. To overcome some of these deficiencies, various attempts have been made to incorporate explicitly plausible dynamic structures into demand systems specifications. More specifically, several alternative hypotheses designed to incorporate into demand systems structures for explaining persistences in consumption behavior patterns have been advanced.

In general, the static models are special or nested versions of their dynamic counterparts because in large part, the dynamic models have been advanced as extensions of specific static representations. An advantage of this relationship between the static and dynamic models is that the likelihood ratio procedure can be used to statistically determine if these more complicated models better capture actual consumer behavior than their static counterparts.[1]

Hypothesized structures for accommodating persistence in consumption behavior are rationalized on the basis of inventories or changes in tastes. In the case of inventories, it is clear that since the consumption is of the flow of services from, say, durable goods, purchase patterns may be quite different from consumption patterns. Also, the existence of inventories of goods, perhaps in the form of durables, can influence the consumption of other goods. An obvious current example is the

Authors of this chapter are Richard Green, Zuhair Hassan, and Stanley R. Johnson.

influence of automobile stocks on the purchase of gasoline at
substantially different relative prices.

Not unrelated to the inventories hypothesis is that of
changing tastes. Clearly, consumer tastes determine the purchases which result in inventories of unconsumed services. Pollack
(1978) has suggested four reasons why tastes in an analysis of
household behavior can be considered to be endogenous: (1)
habit formation, where taste changes are related to past decisions of consumers, (2) interdependent preferences, that is,
preferences related to the consumption patterns of others, (3)
advertising designed explicitly to influence and/or modify the
tastes of individuals, and (4) prices (e.g., "snob appeal"),
whereby people prefer higher priced goods because the price
enhances the appeal.[2]

In this chapter, the primary focus is on alternatives for
the incorporation of persistence via inventory or taste change
hypotheses into demand systems analyses. The dynamic demand
systems developed as extensions of the static models will emerge
in this context. In general, this chapter provides an overview
of the customary ways of incorporating habit formation into
single-equation specifications and demand systems. Following
the general discussion of how habit formation or persistence may
be incorporated into the study of consumption patterns, alternative hypotheses concerning habit formation and stochastic specifications of the disturbance terms are postulated and formulated
into Stone's linear expenditure system for empirical analyses.
Six different versions of the LES models that incorporate alternative hypotheses of habit formation and/or autocorrelation are
estimated and presented in this chapter. The likelihood ratio
test is applied to test the validity of the persistence and autocorrelation hypotheses postulated among the different LES models.
Finally, the empirical results are compared with those of previous studies and the structural implications and significance of
endogenizing persistence through altered structures and various
disturbances processes in demand analyses are discussed.

Specifically, this chapter is organized into the following
five sections: (1) alternative persistence hypotheses, (2) LES
and persistence hypotheses, (3) test of persistence hypotheses,
(4) comparison of results with other studies, and (5) conclusions.

ALTERNATIVE PERSISTENCE HYPOTHESES

A concise overview of some of the more popular changing
tastes formulations is presented here. Persistences in consumption patterns can occur when consumers develop habits for particular commodities or services. By definition, habits are response
patterns to given stimuli (including the cost of acquiring information) that become regularized activities thereby reducing

much of cognitive content of consumption decisions. Hence, consumers tend to rely on and be conditioned by past experiences and alter their behavior only slightly from one period to the next. Also, consumers may be less sensitive to price and income changes when a habit for the commodity has been developed.

Habit formation can be observed by the frequent repetition of purchases of a particular commodity or service. However, since habits are noneconomic entities or at least not endogenized in present consumption theory, they are not as easily quantified or measured as other variables, say, prices and income. To model persistences in consumption analyses, several approaches have been taken. The approaches can be classified into a few general categories.

The first category includes several ad hoc procedures representing different degrees of sophistication. The most simple of these is to add trend variables to the demand equations as derived from the static theory. The objective is to reflect changes in tastes, habits, and other socioeconomic factors through the trend term. However, the underlying theoretical justification for the inclusion of these time variables is lacking. Closely related to this approach is the introduction of trending in the parameters of models based on static theory (Stone 1954). Again, as with the additive trend term, the results obtained by using the method are limited with respect to structural interpretation. That is, while adding such terms may improve the short-run predictive performance of the estimated demand systems, the interpretation of the results is not provided by the overall behavioral hypothesis or model structure. A final procedure included in this category is the "state adjustment" model of Houthakker and Taylor (1970). This model assumes that quantities purchased depend on existing stocks--either physical stocks of goods or psychological stocks or habits. This procedure is again limited in the sense that although it explicitly considers the influence of past consumption behavior on current consumption patterns, it still does not introduce these factors into an overall integrated utility maximizing framework.[3] Thus, an underlying theoretical structure explaining the incorporation of habit formation into the model is still lacking.

The second category to the problem of capturing the observed persistence in consumption patterns in the structure of demand systems follows from the use of a dynamic utility function. More specifically, the utility function is made dynamic by directly incorporating the changing of tastes. The quadratic model of Houthakker and Taylor and the dynamic linear expenditure system of Phlips (1974) are illustrative of this approach. These models are myopic since the current expenditure allocation is assumed to be influenced by past consumption, but the effects of current expenditure allocation on future preferences are ignored. Thus, though the utility function is dynamic, the sequential consumption decision rule implied by the methods used in estimating the parameters is nonoptimal.

The final category of models integrates both past and future considerations into a more coherent treatment of the consumer choice problem. The problem is cast in a control theory format where it is postulated that the consumer is attempting to maximize a discounted utility function subject to wealth and stocks constraints. The stocks constraints can, of course, refer to habits. The models developed by Lluch (1974), Phlips (1974), and Klijn (1977) exemplify these more general attempts to model the sequential consumption decision problem.

The introduction of habit formation hypotheses in demand systems context has progressed along two fronts. Simple structures based on the ideas identified with category two have been combined with systems hypotheses. For example, habit effects have been introduced and tested for in the translog demand system by Manser (1976) and in the linear expenditure by Pollak and Wales (1969), Phlips (1974), and others.[4] Lluch and Williams (1975), MacKinnon (1976), and Green et al. (1978) have provided estimates of the LES assuming autoregressive errors. More recently, Howe et al. (1979), and Green et al. (1980) estimated the LES considering simultaneously the effects of autocorrelation and habit formation hypotheses.

LES MODEL AND PERSISTENCE HYPOTHESES

Habit Formation
As previously developed in Chapter 4, the system of expenditure relations for the linear expenditure system is expressed as

$$p_i q_i = p_i Y_i + \beta_i (y - \sum_{k=1}^{n} p_k Y_k) \quad (i = 1, \ldots, n) \quad (5.1)$$

By allowing the βs and Ys in (5.1) to vary linearly with time and previous consumption levels, q_{t-1} changes in tastes can be incorporated in the specification of the LES.[5] Other variants on incorporating persistence in consumption behavior in a similar ad hoc fashion are of course possible.[6] Within the present context, the structure of habit formation is postulated and written as

$$Y_{it} = Y_i^* + \alpha_i q_{it-1} \quad (5.2)$$

Substituting (5.2) into (5.1) yields a generalized expenditure system $(i = 1, \ldots, n)$ of the form

$$p_{it}q_{it} = [(1 - \beta_i Y_i^*)]p_{it} - \sum_{j \neq i} (\beta_i Y_j^*)p_{jt} + [(1 - \beta_i)a_i]p_{it}q_{it-1}$$

$$- \sum_{j=i} (\beta_i a_j)p_{jt}q_{jt-1} + \beta_i y_t \quad (5.3)$$

where time subscripts have been added to distinguish observations for different periods.

Stochastic Specifications

For empirical implementation, an error term, ε_{it}, is added to (5.3). Due to complications in the estimation process, especially in the presence of lagged endogenous regressors, the disturbances are usually assumed to be time independent. Under these assumptions and using vector notation, we have

$$E(\varepsilon_t) = 0 \quad \text{and} \quad E(\varepsilon_t \varepsilon_t') = \delta_{tt}' \Omega$$

where E is the expectation operator and δ_{tt}' the Kronecker product. That is, the error term has expectation zero, is temporally uncorrelated, and has a contemporaneous variance–covariance matrix Ω. If autocorrelation is assumed in the present context, then

$$\varepsilon_t = R\varepsilon_{t-1} + v_t \tag{5.4}$$

for t = 2, ..., T and where v_2, ..., v_T are independently, identically distributed normal random vectors with mean vector zero and contemporaneous covariance matrix Σ.[7] With the autoregressive assumption in (5.4) and linear habit formulation given by (5.2), the maintained hypothesis has been augmented to accommodate persistence in consumption patterns either structurally or through a process on the disturbances.

Since the sum of individual expenditures equals the total, it follows that the contemporaneous covariance matrix is singular. Barten (1969) has shown that maximum likelihood estimates of the parameters can still be obtained by arbitrarily deleting an equation and that these estimates are invariant to the equation deleted. This result does not hold with autocorrelated disturbances (Berndt and Savin 1975). They show that the adding-up property of the expenditures implies certain restrictions on the autoregressive parameter, R. With no assumed autocorrelation across equations, their results imply that the autocorrelation coefficients must be the same for each equation. With these stochastic assumptions, a program developed by Hall and Hall (1978) and discussed in Berndt et al. (1974) was used to obtain full information maximum likelihood estimates of the parameters of the linear expenditure system.[8]

Empirical Results

Results from the estimation of the LES model can be presented in alternative forms. As implied by the discussion in this section, the structural parameters have an interesting interpre-

tation. Also, the income and price elasticities can be computed. These provide bases for comparisons with results from other
demand systems and applications.

The short-run income and price elasticity formulas obtained
by differentiation of the system of expenditure equations
represented in (5.3) have already been given in Table 4.1 of
Chapter 4.

Parameter estimation. Annual time-series data from Statistics
Canada for the years 1947-72 were employed to estimate the parameters for equations (5.3) and (5.4). The results were in turn
used to compute the implied elasticities and to conduct various
hypotheses tests (for further details see Green et al. 1978).
Consumer expenditure data for three aggregate commodity groups
(durables and semidurables, nondurables, and services) were
used. The commodity group classification was chosen due to the
directly additive utility function associated with the LES and
its restrictive conditions (e.g., of not allowing for
complements). Pragmatically, convergence problems encountered
with using full information maximum likelihood methods
prohibited use of a less aggregated commodity grouping.[9]

A major emphasis of the analysis is testing of restrictions
afforded by the comparisons of models formulated under
alternative habit formation and serial correlation hypotheses.
As a preliminary to the presentation of results for these tests,
the structural parameter estimates and implied elasticities are
examined for six models. The six models are (1) static LES
assuming no autocorrelation, (2) static LES assuming a first-
order autoregressive scheme, (3) LES with a proportional habit
scheme and assuming no autocorrelation, (4) LES with a proportional habit scheme and assuming a first-order autoregressive
error process, (5) LES with a linear habit formation scheme
assuming no autocorrelation, and (6) LES with a linear habit
scheme and assuming a first-order autoregressive error process.
Results from these models provide a basis for the hypothesis
tests on habit persistence, serial correlation, and the
classical consumer demand formulation.

Structural estimates. Only a brief discussion of the estimates
of the structural parameters for the six different versions of
the linear expenditure system is presented. For a more detailed
presentation, see Green et al. (1980). The marginal budget
shares, β_i, are all positive, less than one, and sum to one for
each of the models. They are significantly different from zero
in every case.

The coefficients interpreted as minimum required quantities,
γ_i, are all positive except for durables and semidurables in
the linear habit formation cases. They are not, however,
generally significantly different from zero in the various
models. This result may be symptomatic of multicollinearity

problems arising due to misspecification of persistence
relationships and/or insufficient variation in the sample
data.[10] The presence of lagged consumption is apparently con-
founded with the minimum required quantities effect since in
models 5 and 6 for nondurables and services, the α_i are significant.

For the proportional habit formation systems, models 3 and 4,
the α_i are significantly different from zero for nondurables and
services.   The exception occurs in models 3 and 4 for the durables and semidurables group. This result indicates that persistence in consumption patterns is present in the Canadian data, a
claim that will subsequently be more systematically investigated
using a test for habit effects based on the likelihood ratio.

Estimated marginal budget shares differ little between the
static models, 1 and 2, but the subsistence quantities are
appreciably smaller when the autocorrelation parameter is not
constrained to zero. The autocorrelation coefficient is significantly different from zero in model 2.

In the habit formation versions of the LES, the α_i are higher
for the proportional structures (models 3 and 4) than for the
linear structures (models 5 and 6). When the autocorrelation
parameter is not constrained to zero, it is estimated at .319
with a standard error of .304 for the linear habit model and
.350 with a standard error of .314 with the proportional habit
formation. In both instances the autocorrelation parameter is
not significantly different from zero. Values of the parameters
defining the marginal budget shares are similar for the two
habit formation versions of the LES. The α_i are similar for the
linear habit models (5 and 6) with and without autocorrelation
and for the proportional habit models (3 and 4) with and without
an autocorrelation.

Finally, for the minimum subsistence quantities, the implied
values are less than consumption quantities for every commodity
and time period. Also, for the static versions, the minimum
quantities are larger than consumption for a few commodities for
the first few observations in the time-series. Similar results
were obtained by Pollak and Wales (1969) and are reflective of
the initialization of the model.

Elasticities. Income and price elasticities calculated using
the structural parameters and the sample means are presented in
Table 5.1. Comparing the income elasticities for the six cases
shows that for the nondurables group, quite similar results were
obtained. For the durables and semidurables group, the income
elasticity nearly doubled between the static models (1 and 2)
and the models incorporating a structure for habitual behavior
(3-6). The reverse was true for the services commodity group.

Except for the durables and semidurables commodity group,
the uncompensated direct price elasticities were generally lar-

TABLE 5.1. Income and price elasticity estimates, linear expenditure models 1-6

Commodity group	1	2	3	4	5	6
			Income elasticity			
Durables and semidurables	1.06	1.04	2.32	2.30	2.04	1.96
Nondurables	0.73	0.73	0.60	0.58	0.60	0.63
Services	1.20	1.22	0.38	0.41	0.58	0.62
			Direct price elasticity[a]			
Durables and semidurables	−0.71	−0.77	−0.82	−0.92	−1.01	−1.37
Nondurables	−0.49	−0.60	−0.30	−0.35	−0.44	−0.53
Services	−0.79	−0.91	−0.21	−0.27	−0.44	−0.53

Note: For habit models, short-run elasticities are reported.
[a]Uncompensated.

ger for the static models than for the models with structures admitting persistence. The larger variations in the elasticities are for the services group. The lower price elasticities and low income elasticities were observed for this group when the static structure was modified to allow for persistence in consumption behavior, better reflecting the secular growth in expenditures on this commodity group. Converse but less pronounced results were obtained from the durables and semidurables group where inventories are of obvious importance in influencing expenditure decisions. The most general model (6) showed for the durables and semidurables commodity group a somewhat lower income elasticity and a higher price elasticity than for other models in which the persistence hypothesis was implemented either on the systematic or error structure of the model.

TESTS OF PERSISTENCE HYPOTHESES
    To test for persistence effects, autocorrelation, and the validity of the restrictions implied by the models, a likelihood ratio test was used where

$$\lambda = \max_w L / \max_W L \tag{5.5}$$

The statistic in (5.5) is the ratio of the maximum value of the likelihood function L subject to restrictions to the maximum value of the likelihood function without restrictions. It can be shown that

$$-2 \ln \lambda = T (\ln \Sigma_w - \ln \Sigma_W) \tag{5.6}$$

follows a chi-square distribution, at least asymptotically under the null hypothesis of no persistence effects and/or no autocorrelation. The number of degrees of freedom is equal to the number of restrictions to be tested, where Σ_w is the restricted estimator of the covariance matrix and Σ_W is the unrestricted estimator (Theil 1971).

The approach employed to conduct the tests was to use the general form of the expenditure function (5.3) plus autocorrelation (5.4) as the general hypothesis. Then the null hypotheses tested were (see Fig. 5.1)

H_1: α_i = 0, ρ = 0 (no habit persistence and autocorrelation)

H_2: α_i = 0   (no habit persistence)

H_3: ρ = 0   (no autocorrelation)

If H_1 were rejected, then there would be empirical support for either habit persistence and/or autocorrelation. The next step would consist of testing for habit persistence and autocorrelation. Assuming H_1, H_2, and H_3 were all rejected, the appropriate maintained hypothesis for testing the theoretical restrictions would again be as specified in (5.3) and (5.4).

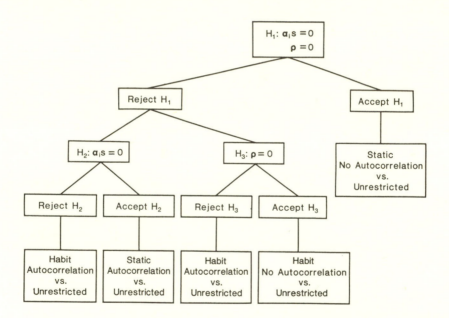

Fig. 5.1. Tests for habits, autocorrelation, and theoretical restrictions.

To test for habit persistence, the likelihood ratio test was applied to models as indicated above. The static models are restricted forms, α_i = 0, of the linear habit persistence model specified in (5.3) and (5.4). The computed value of $-2 \ln \lambda$ was 25.03 for model 6 versus model 2, which is larger than the chi-square value, $\chi^2_{.05,3}$ = 7.815 (note that there are three restric-

tions and, thus, 3 degrees of freedom). Similar results were
obtained for model 5 versus model 1 with $-2 \ln \lambda = 98.01$.
Thus, the null hypothesis was rejected and habit persistence was
assumed present.

To test which structure for habit formation better fitted
the data, the linear habit model, or the proportional habit
model, likelihood ratio tests were again performed. Models 3
and 4 are restricted versions of models 5 and 6, respectively,
where $\Upsilon_i^* = 0$. The value of $-2 \ln \lambda$ of 14.16 for model 6 versus
model 4 is greater than $\chi_{.05,3}^2 = 7.815$, thus indicating a rejection of the null hypothesis (i.e., the linear habit formation
model appeared to describe the data better than the proportional
version). Also a value of $-2 \ln \lambda$ of 14.45 for model 5
versus model 3 was greater than $\chi_{.05,3}^2 = 7.815$, again indicating
ing the same result but with ρ set to zero (i.e., the linear
habit model is preferred over the proportional formulation).

To test whether or not autocorrelation was present, the
likelihood ratio was again used but with one degree of freedom--
it is equivalent to an asymptotic t-test. For the static model
(2) the autocorrelation parameter is significantly different
from zero. When autocorrelation is not constrained to zero in
the linear habit model (6) and the proportional habit model (4),
it is not significantly different from zero. As added evidence,
the more general comparison, model 6 versus model 5, yields -2
$\ln \lambda = 2.44$. Thus, the habit formation structures are apparently
accounting for autocorrelation observed with the static LES within the systematic components of the models.

Finally, two more general tests of the composite habit formation and autocorrelation formulations were run. The comparison
of the static model with no autocorrelation (1) with the most
general formulation (6), gives a very large value for the test
statistic as would be expected given the foregoing results. A
less pronounced difference, although still statistically significant, is obtained by comparing models 3 and 6.

To test for the full set of restrictions implied by each
version for the LES, the models were compared with the unrestricted reduced forms. Degrees of freedom are calculated as
the difference between unrestricted reduced form parameters and
those required to specify the various versions of the LES model.
Note that the "restrictions implied by the theory" are assumed
by the imposition of the LES parameter restrictions. In each
case, the restrictions implied by the theory were not appropriate. These conclusions are not surprising. Many others have
obtained similar results. An implication, however, is that
standard applied theory of consumption may be lacking as a guide
to the specification of aggregate expenditure models even when
incorporating simple structures for reflecting persistence in
consumption behavior either in the structure of the demand
systems or in the disturbance terms.

COMPARISON OF RESULTS WITH OTHER STUDIES

A comparison of our results with other studies is given in Table 5.2. In most cases the autocorrelation coefficient had a high estimated value. The values are -.30, .65, .85, and .99. Thus for the static LES, autocorrelation appears to be present. When habits are introduced into the model structure, the autocorrelation coefficient becomes insignificant in our models. For the Howe et al. model, it was negative and significantly different from zero. Clearly, a more thorough treatment of habit effects and autocorrelation is needed. Sometimes, for example, the inclusion of a time trend works about as well as lagged quantities in describing changes in tastes (Pollak and Wales 1969).

TABLE 5.2. Comparison of results with other static and dynamic LES studies

Study; time period; country	Demand system	Commodity group	Elasticity			
			Price		Income	
			Static	Dynamic	Static	Dynamic
				(in 1961)		
MacKinnon (1976);	LES Static	food	-.48	-.17	0.60	0.29
1927-40, 1948-71;	$\rho = 0$	clothing	-.59	-.66	0.81	1.07
Canada	$\rho \neq 0$	shelter	-.81	-.76	1.07	0.97
	(r = .99)	transportation	-.89	-.77	1.27	1.35
		recreation	-.87	-.89	1.23	1.60
		miscellaneous	-.67	-.50	0.94	0.81
				(at mean values)		
Lluch and Williams	LES Static	food	-.275	--	--	--
(1975);1930-72;	$\rho \neq 0$	clothing	-.478	--	--	--
United States	(r = .85)	housing	-.612	--	--	--
		durables	-.825	--	--	--
		other	-.742	--	--	--
				(at mean values)		
Green et al. (1978);	LES Static	durables and				
1947-72; Canada	$\rho = 0$	semidurables	-.63	-.71	1.07	1.08
	$\rho \neq 0$	nondurables	-.49	-.55	0.73	0.74
	(r = .65)	services	-.74	-.79	1.20	1.17
				(in 1960)		
Howe et al. (1979);	LES Linear	food	-.38	--	--	--
1929-41, 1947-75;	Habits	clothing	-.25	--	--	--
United States	$\rho \neq 0$	shelter	-.17	--	--	--
	(r = -.30)	miscellaneous	-.21	--	--	--

Second, the income and price elasticities obtained in the present analysis appear to be consistent with those from other studies. For example, for durables and semidurables, the estimate of the uncompensated price elasticity was -.71 compared to a value for durables of -.825 found by Lluch and Williams (1975).

CONCLUSIONS

There are several ways of incorporating endogenous tastes into demand systems analyses. The one considered in this chapter was the introduction of a linear habit formation scheme into a demand system. This assumes that tastes depend only on consumption in the previous period or periods.

When autoregressive stochastic specifications are assumed in the presence of habit formation, misspecification errors may be reduced. For example, it is well known that the omission of more complicated autoregressive stochastic structures in dynamic simultaneous systems of relationships can result in serious estimation and prediction problems. To avoid these specification errors a more generalized error structure should be included as part of the maintained hypothesis.

Finally, the various habit schemes and autocorrelation processes can be tested empirically by using likelihood ratio methods. The empirical results presented in this chapter demonstrate that the various estimates are not robust to the different stochastic specifications. Thus, models which employ alternative habit formation schemes should be tested relative to more generalized stochastic specifications.

NOTES

1. Nonnested tests presented, for example, by Cox (1961), Pesaran (1974), and Pesaran and Deaton (1978) can be used to test for different habit schemes if one model is not a restricted version of the other.

2. Stigler and Becker (1977) take the position that all economic analyses should avoid taste formation, change, and differences. They use the household production specification to circumvent some of the problems raised in treating changes in tastes explicitly. However, this extreme viewpoint is strongly challenged by Pollak (1978).

3. They show that, under rather restrictive conditions, the state adjustment model can be derived from a utility maximization process and El-Safty (1976) has shown that it is a special case of a more generalized "learning by doing" model.

4. For excellent treatments of habit formation in demand systems, see Pollak (1976), El-Safty (1976), Hammond (1976), and Pollak (1978).

5. Deaton and Muellbauer (1980) give several reasons why lagged values of dependent variables may appear in demand equations. One of them (and the one used here) is that tastes are affected by previous consumption experiences. However, there still remains the problem of deducing empirically which reason is responsible for generating the lagged responses in the first place.

6. For example, a habit specification could assign declining geometric weights to all past consumption. Mathematically,

$$b_{it} = b_i^* + \alpha_i m_{it-1} \text{ where } m_{it-1} = (1-\delta) \sum_{j=0}^{\infty} \delta^j q_{it-j} \quad (0 < w < 1)$$

(Pollak 1970).

7. More complicated autoregressive-moving average stochastic specifications than the first-order autoregressive process could be treated. These more sophisticated error formulations

can be handled by time-series methods (Box and Jenkins 1970).

8.   The gradient method of Hall and Hall's TSP program was used to obtain maximum likelihood estimates of the parameters of the various models. In a previous paper (Green et al. 1978), a program developed by Wegge (1969a, 1969b) was used; however, for the most general linear habit formulation with autocorrelation convergence was not obtained. Thus, it was decided to use the TSP program with which convergence was attained.

9.   Some of the results similar to those presented have been developed for a four-commodity classification. The estimates for this less aggregated grouping can be obtained from the authors.

10.   The maximum likelihood estimation methods of Wegge (1969a, 1969b) gave somewhat more encouraging results in this regard.

REFERENCES

Barten, A. P.   1969.   Maximum likelihood estimation of a complete system of demand equations. Eur. Econ. Rev. 1:7-73.

Berndt, E. R., B. H. Hall, R. C. Hall, and J. A. Hausman.   1974. Estimation and inference in nonlinear structural models. An. Econ. Soc. Meas. 3/4:653-65.

Berndt, E. R., and N. E. Savin.   1975.   Estimation and hypothesis testing in singular equation systems with autoregressive disturbances. Econometrica 43:937-57.

Box, G. E. P., and G. Jenkins.   1970.   Time series analysis, forecasting and control. San Francisco: Holden Day.

Cox, D. R.   1961.   Tests of separate families of hypotheses. Proc. Fourth Berkeley Symp.

Deaton, A., and J. Muellbauer.   1980.   Economics and consumer behavior. Cambridge: Cambridge Univ. Press.

El-Safty, A.   1976.   Adaptive behavior, demand and preferences. J. Econ. Theory 13:229-318.

Green, R., Z. Hassan, and S. R. Johnson.   1978.   Maximum likelihood estimation of linear expenditure systems with serially correlated errors: An application. Eur. Econ. Rev. 11:207-19.

_____.   1980.   Testing for habit formation, autocorrelation and theoretical restrictions in linear expenditure systems. South. Econ. J. 47:433-43.

Hall, B., and R. Hall.   1978.   Time series processor. Version 3.3 for the Burroughs 6700/7700 (User's manual).

Hammond, P.   1976.   Endogenous tastes and stable long-run choice. J. Econ. Theory 13:329-40.

Houthakker, H. S., and L. D. Taylor.   1970.   Consumer demand in the United States 1929-1970. 2d ed. Cambridge, MA: Harvard Univ. Press.

Howe, H., R. Pollak, and T. J. Wales.   1979.   Theory and time series estimation of the quadratic expenditure system. Econometrica 47:1231-48.

Klijn, N. 1977. Expenditure, savings and habit formation: A comment. Int. Econ. Rev. 18:771–78.

Lluch, C. 1974. Expenditure, savings and habit formation. Int. Econ. Rev. 15:786–97.

_____, and R. Williams. 1975. Consumer demand systems and aggregate consumption in the U.S.A: An application of the extended linear expenditure system. Can. J. Econ. 8:49–66.

MacKinnon, J. 1976. Estimating the linear expenditure system and its generalizations. In Studies in nonlinear estimation, eds. S. Goldfield and R. Quandt. Cambridge: Ballinger.

Manser, M. 1976. Elasticities of demand for food: An analysis using nonadditive utility functions allowing for habit formation. S. Econ. J. 43:879–91.

Pesaran, M. 1974. On the general problem of model selection. Rev. Econ. Stud. 41:153–70.

_____, and A. Deaton. 1978. Testing non-nested nonlinear regression models. Econometrica 46:677–94.

Phlips, L. 1974. Applied consumption analysis. Amsterdam: North-Holland.

Pollak, R. A. 1970. Habit formation and dynamic demand functions. J. Polit. Econ. 78:745–62.

_____. 1976. Habit formation and long-run utility functions. J. Econ. Theory 13:272–97.

_____. 1978. Endogenous tastes in demand and welfare analysis. Discuss. Pap. No. 393, Univ. of Pennsylvania, Dep. of Econ.

_____, and T. J. Wales. 1969. Estimation of the linear expenditure system. Econometrica 37:611–28.

Statistics Canada. 1975. National income and expenditures accounts. Cat. 13-531, Ottawa.

Stigler, G., and G. S. Becker. 1977. De gustibus non est disputandum. Am. Econ. Rev. 67:76–90.

Stone, R. 1954. Linear expenditure systems and demand analysis: An application to the patterns of British demand. Econ. J. 64:511–27.

Theil, H. 1971. Principles of econometrics. New York: Wiley.

Wegge, L. 1969a. A family of functional iterations and the solution of maximum likelihood estimating equations. Econometrica 37:122–30.

_____. 1969b. Full information quasi-maximum likelihood estimators. Discuss. Pap. Univ. of California, Davis, Econ. Dep.

CHAPTER 6

Analysis of Household Demand for Meat, Poultry, and Seafood Using the S_1-Branch System

RAMPANT INCREASES in the prices of food and nonfood items, dramatic changes in lifestyles, and salient changes in sociodemographic characteristics have distinguished to some extent the 1970s and, thus far, the 1980s from past decades. Increases in the prices of energy, housing, medical supplies and services, and durable goods have raised issues concerning their effects upon the demand for food. Installments due on consumer debt for durable and semidurable goods and other fixed commitments such as house payments (rent or mortgage), taxes, insurance, transportation, and utility bills have a first lien on household income. These commitments have risen sharply over the past few years, and food expenditures and other categories of living expenses may be affected by this trend.

The large-scale admission of women into the labor force and the affluence of U.S. households impact on the demand for food away from home and for prepared foods (Prochaska and Schrimper 1973; Redman 1980). Sociodemographic forces, particularly the size and composition of households, region, and degree of urbanization may exert a notable influence on food consumption. This hypothesis is primarily attributable to the shifts in the response of consumption to the life cycle, differences in the accessibility of the product, differences in climate, and the development of buying habits.

Ameliorating decision making for the food industry requires reliable measures of the impacts of changes in food and nonfood prices, changes in income, and changes in sociodemographic variates. These measures help to explain current economic conditions in the food industry and to forecast future economic conditions.

In this light, the aim of this chapter is to identify and

Authors of this chapter are Oral Capps, Jr., and Joseph Havlicek, Jr.

assess selected factors that affect the household demand for
meat, poultry, and seafood both nationally and regionally.
Special attention is given to meat, poultry, and seafood due to
their relative importance in retail consumption. These items
generally account for 25 to 30 percent of each food dollar in
the household budget (Smallwood and Blaylock 1981). By providing a conceptual framework to deal with the interdependency of
demand for various commodities, the complete systems approach is
very effective for modeling consumption behavior (Barten 1977).
Consequently, a system of demand equations is specified and
estimated.[1] The first and second sections deal with the empirical model and sociodemographic characteristics, while the third
and fourth sections deal with the data base and the estimation
procedures. The fifth section contains the statistical results
and the economic analyses, and the sixth concerns the evaluation
of the empirical model. Concluding comments follow in the last
section.

EMPIRICAL MODEL

Empirical application requires the specification of the
functional form of the demand equations. A generalized linear
expenditure system proposed by Brown and Heien (1972), the
S_1-branch system, is employed.[2] The S_1-branch system is derived
from the specification of a direct utility function (see Brown
and Heien). Consequently, the demand equations of the S_1-branch
system automatically satisfy the classical constraints of demand
theory for any set of coordinates.[3] In contrast, as mentioned
previously in Chapter 1, some researchers have preferred to
work with directly specified demand systems, imposing the constraints which insure their theoretical plausibility. However,
in this case the classical restrictions can only be enforced at
some local set of coordinates.

The S_1-branch system possesses properties more general than
the widely estimated linear expenditure system (Chapters 4 and
5). In addition, the demand system permits a fine classification of commodities where other systems allow only broad categories of items. Consider a partition of a set of n commodities
$(q^1, q^2, \ldots, q^n)$ into S sets of branches, $Q^1, Q^2, \ldots, Q^S$. Let
q^{ri} denote the quantity of the ith commodity in the rth branch
Q^r, and let n^r denote the number of commodities in the rth branch
such that $\Sigma_{r=1}^{S} n^r = n$. The union of the S sets is equivalent to
the set $(q_1, q_2, \ldots, q_n)$ and the intersection of any two distinct branches is the null or empty set. In essence, the S sets,
$Q_1, Q_2, \ldots, Q_S$, are mutually exclusive and exhaustive.

The S_1-branch system corresponds to the Strotz-Gorman utility tree concept. The model constitutes a strongly Pearce–separable demand system, strongly separable among groups and strongly separable among goods within groups (Barten 1977). However, it is important to note that strong Pearce–separability does not imply overall strong separability. Under overall strong separability, as pointed out by Deaton (1978), there exists the unduly restrictive proportionality of the own–price elasticity and the expenditure elasticity of a commodity. Under strong Pearce-separability, more flexibility is permitted (Barten 1977, 40).

Mathematically, the demand relationships of the S_1-branch system are

$$q_{ri} = Y_{ri} + \left(\frac{\beta_{ri}}{p_{ri}}\right)^{\sigma_r} \left[\sum_{i=1}^{n_r} \left(\frac{\beta_{ri}}{p_{ri}}\right)^{\sigma_r} p_{ri}\right]^{-1} w_r \left(y - \sum_{r=1}^{S} \sum_{i=1}^{n_r} p_{ri} Y_{ri}\right)$$

(6.1)

for all $i = 1, 2, \ldots, n_r$ and $r = 1, 2, \ldots, S$, where q_{ri} = the quantity of the ith commodity in the rth branch Q_r, p_{ri} = the price of the ith commodity in the rth branch Q_r, and y = the total expenditure on the vector of commodities. In these respective relationships, Y_{ri} is a "subsistence," "threshold," or "committed" quantity parameter for goods i within Q_r.

The parameter β_{ri} reflects the importance of particular commodities within branches in the generation of utility. For this reason, β_{ri} is labeled a "commodity-intensity" parameter. The branch elasticities of substitution, the parameters σ_1, σ_2, $\ldots$, σ_S indicate the degree of substitutability among commodities in the particular branch of interest. The parameter w_r denotes the share of supernumerary expenditure,

$$y - \sum_{r=1}^{S} \sum_{i=1}^{n_r} p_{ri} Y_{ri},$$

allocated to the supernumerary commodities, $q_{ri} - Y_{ri}$, in branch Q_r.

The following restrictions hold:

$\beta_{ri} > 0$, $q_{ri} - Y_{ri} > 0$ for all i, r, w_r

$\sigma_r > 0$ for all r

$\sum_{r=1}^{S} w_r = 1$, where $i = 1, 2, \ldots, n_r$, and $r = 1, 2, \ldots, S$.

In the Hicks-Allen sense, all intergroup pairs of goods in this system are substitutes, but intragroup pairs of goods may be either substitutes or complements. The expenditure elasticities are unequivocally positive, and thus inferior goods are inadmissible. The ordinary own-price elasticities are nonpositive, assuming any value from minus infinity to zero. Hence, the Giffen paradox may not occur, but elastic and inelastic demands are permissible.

SOCIODEMOGRAPHIC CHARACTERISTICS

Sociodemographic characteristics have been rarely used as arguments in complete systems of demand equations. As described in Chapter 1, these variates can be introduced into any complete demand system in two ways; this study employs the translating technique of Pollak and Wales (1978). The chosen characteristics were household size and degree of urbanization. Degree of urbanization, or population density, refers to Standard Metropolitan Statistical Areas (SMSAs), classified as follows: 1,000,000 and over population, 400,000 to 999,999 population, 50,000 to 399,999 population, and urban and rural areas outside SMSAs.

It is recognized that the treatment of sociodemographic factors is limited. Sociodemographic characteristics such as ethnic background, employment status, education, and household age-sex composition are certainly relevant and hence also warrant consideration. However, incorporating sociodemographic factors into demand systems via the translating method substantially increases the number of parameters to be estimated. To ease the computational burden, only household size and degree of urbanization are considered. Price (1970) and Buse and Salathe (1978) have shown that the impact of additional members on food expenditures decreases with increases in household size. Consequently, the square of household size is included as an additional explanatory variable in the demand relationships.

It is assumed that the subsistence parameters Y_{ri} depend linearly on degree of urbanization, household size, and the square of household size. Mathematically,

$$Y_{ri} = v_{0ri} + v_{1ri}D_1 + v_{2ri}D_2 + v_{3ri}D_3 + v_{4ri}HS + v_{5r}HS^2 \quad (6.2)$$

where D_1 = 1 SMSA with population 1,000,000 and over, 0 otherwise; D_2 = 1 SMSA with population 400,000 to 999,999, 0 otherwise; D_3 = 1 SMSA with population 50,000 to 399,999; 0 otherwise; and HS = household size.

Urban and rural areas outside SMSAs is the base or omitted category for the intercept shifters, D_1, D_2, and D_3. Substituting (6.2) into (6.1) generates the demand functions to be estimated. Additionally, it is important to note that by making the

subsistence parameters functions of selected household characteristics, the rigidity of this demand system, in potentially
not allowing for nontrivial interactions among commodities due
to the strong separability condition within the branches, is
partially relaxed. In similar fashion, to approach the issue of
rigidity in this demand system, Deaton (1974) made the subsistence parameters functions of relative prices.

The enumeration of the branches and the respective commodities are: branch 1--ground beef, roasts, steaks, pork, other
meats, poultry, and seafood; branch 2--other foods and food away
from home; and branch 3--fuels for home heating and gasoline.
These branch categories are formulated with the particular purpose of analyzing the impacts of nonfoods, food away from home,
and other foods on the demand for meat, poultry, and seafood.
The second branch is constituted of all food categories (either
at home or away from home) other than meat, poultry, and seafood.   The third branch deals with nonfood (energy) commodities. Data for other nonfood items such as housing, medical
services, durables, and clothing were not available from the
source of data, the 1972-74 U.S. Bureau of Labor Statistics
Consumer Expenditure Diary Survey (BLS CEDS). In sum, the
partition of the 11 commodities into the 3 distinct separable
branches permits concentration on the demand for meat, poultry,
and seafood, taking into account additional factors.

The assignment of commodities into separable groups was
primarily intuitive in the absence of a priori knowledge of
actual relationships among marginal utilities. Obviously, other
partitions of the commodity set exist, but this particular partition has appeal on heuristic grounds. Due to the fact that
the sum of the expenditures on the individual goods equals the
total expenditure under consideration, this set of consumption
categories constitutes a complete demand system.

DATA BASE

The BLS CEDS provides a source of expenditure, quantity, and
income information in relation to sociodemographic characteristics of over 20,000 households in the United States. These
households kept a record of purchases of food and particular
nonfood items for two consecutive weeks.

The sample in this study included (1) households that reported the sociodemographic characteristics of household size and
degree of urbanization by region, (2) households that reported
purchase information of at least one meat, poultry, or seafood
commodity by region, and (3) households that reported purchase
information on food away from home, other foods, fuels for home
heating, and gasoline by region.

This study explores both national and regional household
demand relationships. The regions are the United States, the
Northeast, the North Central, the South, and the West. The
number of observations in the sample for each region is 4,041
observations for the United States, 855 observations for the

Northeast, 1,293 observations for the North Central, 1,180 observations for the South, and 713 observations for the West. Moreover, the size of the samples permits the use of cross validation to assist in evaluating the empirical performance of the
S_1-branch system. Since the use of cross validation requires the
availability of independent samples, the respective samples were
split into two random samples of almost equal size for each
region.[4] Prices for meat, poultry, and seafood were derived from
expenditure and quantity data. Regional price indices were employed for gasoline, fuels for home heating, food away from home,
and other foods due to the unavailability of quantity data for
these commodities. The variation in the derived prices was
greater by far than the variation in the regional price indices.

With reference to meat, poultry, and seafood, the problems
of how to handle sample observations with zero-level consumption
came to light in this application. From the BLS CEDS it was
impossible to determine whether the household did not purchase
the particular products at all or simply did not make the purchases during the consecutive two-week period. Because the
survey period for the household was short, nonpurchases may
reflect that the household consumed out of inventories.

The solution to the zero-level consumption problem depends
on the reason the household does not purchase particular
products. This study favors retaining sample observations with
zero expenditure and consumption levels to adequately portray
the full range of observed behavior. As indicated in Table A6.1
at the end of this chapter, the percentage of zero expenditures
for roasts and steaks was approximately 50 percent, the percentage of zero expenditures for ground beef, poultry, and seafood
was about 40 percent, and the percentage of zero expenditures
for pork and other meats was roughly 20 percent.

The prices associated with these zero expenditures were the
maximum observed sample prices increased by 10 percent. The
implicit assumption made in adopting this solution is that households do not enter the market once their threshold price is
surpassed. Although somewhat artificial, this threshold price
concept was employed because it was not possible to estimate the
S_1-branch system using only those households purchasing all meat,
poultry, and seafood products. The number of observations would
have been substantially less than the number of parameters to be
estimated. Moreover, although the use of artificial data generally has serious implications for parameter estimation, the parameter estimates in this study were particularly quite robust
with respect to various price levels associated with zero expenditures, ranging from the sample mean to 20 percent above the
maximum sample price. The mean prices and quantities of meat,
poultry, and seafood by region and by sample are exhibited in
Table A6.2 at the end of this chapter. To take account of the
zero-expenditure problem, the possibility of generalizing Tobit
analysis to the estimation of the demand system was

considered.However, because the problem of what price levels to
associate with zero expenditures still remained, this complex
procedure was not undertaken.

ESTIMATION PROCEDURE

For empirical implementation, any demand system model must
be embedded in a stochastic framework. A disturbance term for
each equation in the system is required since some factors not
explicitly introduced in the model may influence household consumption behavior. It is assumed that the disturbances enter
each equation of the S_1-branch system in an additive fashion.

Let e_{ri} denote the disturbance term of the ith demand equation
in the rth branch. The assumptions of the error structure are
as follows: (1) $E(e_{ri}) = 0$ and (2) $E(e_{ri}e_{tj}) = \delta_{ri,tj}$. Denote the
variance-covariance matrix of the disturbance terms as $\Omega =$
$(\delta_{ri,tj})$. The covariances between the disturbance terms for any
two goods in the same branch or in two different branches is constant. The stochastic specification allows nonzero contemporaneous covariances both among errors for commodities in the same
branch and among disturbance terms for commodities from different branches. This specification rules out nonzero noncontempor-
aneous covariances of the disturbance terms. In short, the
disturbance terms, by assumption, have the classical properties.

Due to the adding-up condition, the variance-covariance
matrix of the disturbance terms is singular. Each disturbance
term can be written as a linear combination of the remaining
disturbance terms. The singularity of Ω prohibits the estimation of the demand functions by full system approaches. To
overcome this singularity, it is necessary to delete arbitrarily
one commodity from the full set.

To take into account parameter nonlinearity and cross-equation correlation, a full-information-maximum likelihood (FIML)
algorithm was employed (Bard 1967). Most practitioners use this
type of computation routine to handle nonlinear systems of equations under the assumption that the disturbance terms are joint-
ly normally distributed. Defining $\widetilde{\Omega}^{ri}$ as the variance-covariance matrix of disturbance terms with e_{ri} omitted, the Bard
algorithm maximizes the likelihood function associated with the
remaining commodities.[5] The FIML procedure produces parameter
estimates invariant to the choice of deleted commodity (Berndt
and Savin 1975). The omitted commodity in this study was gasoline.

STATISTICAL RESULTS AND ECONOMIC ANALYSES

Estimates were obtained of system parameters and own-price,
cross-price, expenditure, and household-size elasticities for

regions of the United States and the United States as a whole. The estimates of parameters and elasticities from the S_1-branch system were reasonable on a priori grounds. Maximum likelihood estimates and standard errors of the parameters in the S_1-branch system for the United States (sample 2) are exhibited in Table 6.1.[6] This set of parameter estimates and standard errors is representative of the parameter estimates and standard errors for other regions and other samples. For any region and sample, most of the estimated parameters were statistically different from zero at any reasonable level of significance.

For each sample and region, the ß, σ, and w were all positive and significantly different from zero. The consumption of roasts, steaks, and pork were relatively more important in generating utility than ground beef, other meats, poultry, and seafood. As expected, the degree of substitutability among commodities in the various branches was highest in the meat, poultry, and seafood branch, next highest in the other foods and food away from home branch, and lowest in the nonfoods (energy) branch. The share of the supernumerary budget expenditure allocated to the supernumerary commodities in the second and third branches was approximately 38 percent. The share allocated to the meat, poultry, and seafood branch was roughly 24 percent.

The expressions for the respective elasticity measures have been derived by Brown and Heien (1972, 742). Although the patterns were similar, a different set of estimates was associated with each region. Due to space limitations only the own-price and expenditure elasticities are shown in Table 6.2. The estimated elasticities and the variances of the sampling distributions of the elasticities involve complex nonlinear transformations of the estimated parameters. Hence, the respective variances have not been explicitly derived, and the investigation of their statistical reliability through formal tests of significance is precluded.

All the own-price coefficients were negative for the respective commodities in conjunction with theoretical expectations. In particular, the own-price elasticities for meat, poultry, and seafood exceeded unity. Such elastic responses to own-price changes may be attributable in part to the level of disaggregation of the commodities and to the type of data. Estimates based on household survey data typically represent longer-run behavior than estimates based on time-series data (Kuh 1959). The level of disaggregation of meat increases the number of substitutable products.

The demand for ground beef and pork in the Northeast was more elastic than in the North Central, South, and West. The demand for steaks in the North Central and the South was more elastic than in the Northeast and West. Further, the demand for seafood was more elastic in the South than in the other regions.

TABLE 6.1. Maximum likelihood estimates and standard errors of parameters, S_1-branch system for United States (Sample 2)

Parameter		Commodity coefficients associated with demographic characteristic (τ)			
Commodity-intensity (β)[a]					
β_{11}	.3389[b] (.0070)[c]	τ_{011}	−.9073 (.0912)	τ_{314}	−.5217 (.0111)
β_{12}	.5193 (.0028)	τ_{111}	.0215 (.0103)	τ_{414}	.2244 (.0184)
β_{13}	.5572 (.0093)	τ_{211}	.1146 (.0948)	τ_{514}	−.0080 (.0022)
β_{14}	.6103 (.0108)	τ_{311}	−.0489 (.0377)	τ_{015}	.4817 (.0315)
β_{15}	.3782 (.0046)	τ_{411}	.1286 (.0110)	τ_{115}	.2719 (.0346)
β_{16}	.3047 (.0093)	τ_{511}	−.0047 (.0013)	τ_{215}	−.0913 (.0412)
β_{17}	.3068 (.0102)	τ_{012}	−.8117 (.0736)	τ_{315}	−.2079 (.0315)
β_{21}	.7923 (.0128)	τ_{112}	.1478 (.0451)	τ_{415}	.0982 (.0053)
β_{22}	.8647 (.0132)	τ_{212}	.0927 (.0512)	τ_{515}	−.0018 (.0007)
β_{31}	.9927 (.0027)	τ_{312}	.2693 (.0778)	τ_{016}	−.5121 (.0519)
β_{32}	.3385 (.0067)	τ_{412}	.0883 (.0113)	τ_{116}	−.0177 (.0091)
Branch elasticities of substitution (σ)[d]		τ_{512}	−.0032 (.0011)	τ_{216}	−.2912 (.0612)
σ_1	1.1727 (.0105)	τ_{013}	−.4817 (.0428)	τ_{316}	−.3741 (.0441)
σ_2	.9748 (.0142)	τ_{113}	−.1692 (.0391)	τ_{416}	.0335 (.0053)
σ_3	.5839 (.0155)	τ_{213}	−.0412 (.0396)	τ_{516}	−.0017 (.0011)
Supernumerary budget shares (w)[d]		τ_{313}	.2918 (.0312)	τ_{017}	−1.2916 (.0712)
w_1	.2382 (.0127)	τ_{413}	.0827 (.0131)	τ_{117}	.5199 (.0612)
w_2	.3888 (.0201)	τ_{513}	−.0030 (.0012)	τ_{217}	−.0311 (.0198)
w_3	.3730 (.0154)	τ_{014}	−.2392 (.0337)	τ_{317}	−.0691 (.0178)
		τ_{114}	−.6912 (.0675)	τ_{417}	.2524 (.0850)
		τ_{214}	−.9812 (.0079)	τ_{517}	−.0070 (.0087)

Note: Only parameter estimates for the ν_i associated with meat, poultry, and seafood are presented here due to space constraints. For each region and sample, 83 parameters were estimated simultaneously.
[a]Subscripts for parameters indicate commodity in question: 11–ground beef, 12–roasts, 13–steaks, 14–pork, 15–other meats, 16–poultry, 17–seafood, 21–other foods, 22–food away from home, 31–fuels for home heating, 32–gasoline.
[b]Parameter estimate.
[c]Standard error.
[d]Subscripts for parameters indicate the branch categories.

The differences in own-price elasticities among the regions for roasts, poultry, and other meats for the most part was negligible.

The commodities that were most responsive to own-price

TABLE 6.2. Own-price elasticities and expenditure elasticities

Commodity	United States S1[a]	United States S2[a]	Northeast S1	Northeast S2	North Central S1	North Central S2	South S1	South S2	West S1	West S2
					Own-price elasticities					
Ground beef	-1.58	-1.51	-1.79	-1.76	-1.52	-1.38	-1.63	-1.60	-1.37	-1.31
Roasts	-1.83	-1.83	-1.68	-1.87	-2.01	-1.81	-1.60	-1.82	-2.06	-1.89
Steaks	-1.69	-1.75	-1.44	-1.74	-1.82	-1.87	-1.88	-1.63	-1.48	-1.77
Pork	-1.30	-1.52	-1.72	-1.47	-1.14	-1.45	-1.57	-1.51	-0.67	-1.73
Other meats	-1.46	-1.30	-1.55	-1.42	-1.42	-1.37	-1.50	-0.97	-1.39	-1.64
Poultry	-1.25	-1.28	-1.26	-1.25	-1.17	-1.28	-1.32	-1.26	-1.28	-1.38
Seafood	-2.24	-2.02	-2.06	-1.77	-2.54	-1.59	-2.59	-2.80	-1.37	-1.81
					Expenditure elasticities					
Ground beef	1.38	1.16	1.40	1.27	1.49	1.11	1.21	1.32	1.45	0.90
Roasts	1.66	1.44	1.34	1.40	1.99	1.47	1.20	1.52	2.21	1.30
Steaks	1.51	1.38	1.14	1.29	1.80	1.54	1.44	1.36	1.56	1.26
Pork	1.11	1.05	1.38	1.09	1.11	1.19	1.20	0.79	0.70	1.22
Other meats	1.28	1.13	1.21	1.03	1.39	1.09	1.11	1.27	1.48	1.11
Poultry	1.10	1.00	0.99	0.92	1.13	1.04	1.00	1.05	1.34	0.97
Seafood	1.96	1.56	1.60	1.26	2.52	1.26	1.92	2.32	1.47	1.21

Note: Estimates hold at sample mean quantities, prices, and average budget shares.
Substantial movement away from such coordinates may involve dramatic changes in economic
indices.
[a]Sample 1 and sample 2.

changes (ground beef, steaks, roasts, and seafood) were also
most responsive to expenditure changes. In general, the expenditure elasticities for roasts, steaks, and seafood in the North
Central were larger than in the Northeast, South, and West. The
differences in expenditure elasticities among the various regions for ground beef, poultry, pork, and other meats were not
typically salient.

     In all regions, the influence of substitutes and complements
was for the most part weak. Meat, poultry, and seafood were
substitutes, with interchanges probably occurring for the sake
of variety and economy. Food consumed away from home was a
substitute for ground beef, steaks, roasts, poultry, pork, and
other meats consumed at home. On the other hand, food consumed
away from home was a complement for seafood. Other foods were
substitutes for meat, poultry, and seafood. However, the consumption of meat, poultry, and seafood were independent of
changes in the prices of fuels for home heating and gasoline.

     The coefficients of the number of household members were
positive and at least twice their standard errors for meat,
poultry, and seafood. Household purchases of pork and seafood
were the most sensitive to changes in household size. The
coefficients associated with the square of the number of household members were negative for meat, poultry, and seafood, and
except for poultry and seafood, were at least twice their standard errors, indicating the presence of economies of household
size.

     The influence of degree of urbanization on household consump-
tion of meat, poultry, and seafood was not clear. Although
population density affected these household consumption patterns,
the trends were not uniform across the United States. In the

Northeast and in the North Central, households residing within
SMSAs typically purchased significantly more pork, poultry, and
seafood and significantly less ground beef and other meats than
households residing outside SMSAs. Moreover, in the South,
households residing within SMSAs generally purchased significantly more ground beef and roasts and significantly less steaks,
pork, and poultry than households residing outside SMSAs. In
the West, households located within SMSAs generally consumed
more pork and less ground beef and other meats than households
located outside SMSAs.

EVALUATION OF EMPIRICAL MODEL
     The evaluation of the S_1-branch system rests on (1) use of
parameter values, signs, magnitudes, and test statistics to provide information about the validity of the results, (2) goodness-
of-fit to sample data, and (3) predictive ability to independent
data samples. Probably the most common criterion for model evaluation complementary to examination of theoretical and statistical support of parameter estimates involves goodness-of-fit.
The measure describing the goodness-of-fit is Theil and Mnookin's
(1966) information inaccuracy statistic (IIS).
     Mathematically,

$$IIS = \sum_{r=1}^{S} \sum_{i=1}^{n_r} w_{ri}^{**} \ln(w_{ri}^{**} / \hat{w}_{ri}^{**}), \qquad (6.3)$$

where w_{ri}^{**} denotes the observed average budget share of the ith
commodity in the rth branch and $\hat{w}_{ri}^{**}$ denotes the predicted average
budget share of the ith commodity in the rth branch. Information
inaccuracy values close to zero indicate exceptional fits of the
demand system to the sample data. The information inaccuracy
statistics for the S_1-branch system by sample and region varied
from .0015 to .0071.
     The acid test concerns the predictive performance on the
basis of independent samples. With the availability of independent samples, the evaluation of predictive ability rests on
cross validation. One sample, called the "estimation data," is
used to estimate the coefficients in the demand system. The
other sample, called the "prediction data," is used to measure
the predictive accuracy of the demand system based on the previously estimated coefficients. Pearson product-moment correlation coefficients, median absolute residuals, mean absolute
residuals, and mean squared error of observed and predicted
purchases typically serve as statistical measures of predictive
ability.
     The Pearson product-moment correlation coefficients were
comparatively large for ground beef, roasts, steaks, pork, and

poultry (Table 6.3). In all cases, actual and predicted purchases of the commodities were positively associated. Although
not shown, the mean and median absolute residuals as well as the
mean squared error of observed and predicted purchases were relatively small for meat, poultry, and seafood. Thus, on the
basis of cross validation, the predictive performance of the
S_1-branch system was particularly noteworthy.

TABLE 6.3. Pearson product-moment correlation coefficients of observed and
predicted quantities

Commodity	Correlation coefficients[a]									
	United States		Northeast		North Central		South		West	
	S1[b]	S2[b]	S1	S2	S1	S2	S1	S2	S1	S2
Ground beef	.76	.71	.78	.72	.77	.68	.77	.76	.82	.72
Roasts	.79	.79	.83	.74	.79	.84	.78	.76	.77	.79
Steaks	.82	.79	.80	.77	.83	.80	.77	.84	.88	.74
Pork	.74	.32	.73	.80	.78	.63	.76	.71	.70	.39
Other meats	.57	.53	.64	.60	.57	.56	.39	.48	.64	.43
Poultry	.70	.71	.65	.65	.69	.69	.70	.72	.71	.75
Seafood	.36	.42	.40	.43	.29	.38	.54	.47	.33	.35

[a]All statistically significant at .10 level.
[b]Sample 1 and sample 2.

CONCLUDING COMMENTS
     Brown and Heien's S_1-branch system was employed to investi-
gate the national and regional demand patterns for meat, poultry,
and seafood in the United States. Meat, poultry, and seafood
purchases were very sensitive to own-price changes, changes in
total expenditure, and changes in household size, and were less
sensitive to changes in prices of substitutes and complements,
regional differences, and differences in degree of urbanization.
Energy price changes had hardly any influence on the consumption
of meat, poultry, and seafood.
     Regarding structural estimation, the estimates of parameters
and elasticities in the S_1-branch system were reasonable on a
priori grounds. Also, the goodness-of-fit of the S_1-branch
system for each region and sample was exceptional. Finally,
using independent samples, the predictive performance of the
S_1-branch system was particularly notable.

     To better analyze household consumption patterns of meat,
poultry, and seafood, a more refined disaggregation of these
commodities may be worthwhile. Further, since nonfood and food
prices as well as expenditures have risen sharply since the
period 1972-74, more perspective may be gained with regard to
household purchases of meat, poultry, and seafood in response to
changes in prices and expenditures through the use of data samples for the period 1975-84. In addition, in focusing on region,
household size, and degree of urbanization, other sociodemographic factors are neglected. Economists must be attuned to

changes in a number of sociodemographic variates. Additional efforts should provide more insight in analyzing household consumption patterns of meat, poultry, and seafood.

APPENDIX

TABLE A6.1. Percentage of zero expenditures for meat, poultry, and seafood by region and sample

Commodity	United States		Northeast		North Central		South		West	
	S1[a]	S2[a]	S1	S2	S1	S2	S1	S2	S1	S2
Ground beef	39.43	37.32	46.93	46.40	35.21	34.37	39.75	39.39	37.77	28.04
Roasts	55.17	56.38	44.81	44.54	56.85	56.20	56.62	60.93	61.94	63.45
Steaks	54.03	52.62	50.00	47.23	55.18	55.10	57.31	52.58	51.38	54.67
Pork	19.09	17.67	21.69	22.04	17.68	13.81	14.11	14.69	26.66	24.36
Poultry	41.86	39.25	37.26	38.51	44.35	42.85	36.83	34.89	50.83	41.07
Other meats	18.55	17.52	16.03	18.09	15.39	11.93	20.65	18.53	23.88	25.21
Seafood	44.13	40.89	37.73	41.06	46.03	40.65	44.06	42.57	48.33	38.24

[a]Sample 1 and sample 2.

TABLE A6.2. Mean prices and quantities of meat, poultry, and seafood by region and sample

Commodity	United States		Northeast		North Central		South		West	
	S1[a]	S2[a]	S1	S2	S1	S2	S1	S2	S1	S2
Ground beef	1.49[b]	1.46	1.62	1.62	1.45	1.43	1.50	1.46	1.39	1.32
	(2.52)[c]	(2.65)	(1.77)	(2.02)	(2.97)	(2.99)	(2.23)	(2.40)	(3.04)	(3.22)
Roasts	2.20	2.18	2.01	1.93	2.22	2.20	2.23	2.26	2.31	2.33
	(2.21)	(2.18)	(2.79)	(3.40)	(2.12)	(2.05)	(2.00)	(1.82)	(2.03)	(1.57)
Steaks	2.66	2.68	2.53	2.49	2.72	2.80	2.76	2.70	2.54	2.68
	(2.15)	(2.14)	(2.32)	(2.44)	(1.98)	(1.85)	(1.92)	(2.02)	(2.62)	(2.50)
Pork	1.44	1.43	1.47	1.48	1.45	1.42	1.35	1.35	1.53	1.52
	(4.97)	(4.96)	(4.73)	(5.02)	(5.15)	(5.21)	(5.16)	(5.26)	(4.60)	(3.95)
Other meats	0.97	0.97	0.98	0.98	0.97	0.97	0.98	0.97	0.95	0.97
	(3.43)	(3.53)	(4.22)	(5.02)	(3.55)	(3.50)	(2.87)	(2.77)	(3.20)	(3.04)
Poultry	1.86	1.80	1.69	1.70	2.00	2.00	1.73	1.66	2.03	1.82
	(4.15)	(4.36)	(4.66)	(5.23)	(3.65)	(3.78)	(4.53)	(4.47)	(3.84)	(4.15)
Seafood	1.27	1.27	1.28	1.28	1.27	1.27	1.26	1.27	1.25	1.27
	(1.15)	(1.20)	(1.45)	(1.40)	(0.92)	(0.89)	(1.15)	(1.33)	(1.20)	(1.27)

[a]Sample 1 and sample 2.
[b]Mean price ($/pound).
[c]Mean quantity (pounds).

NOTES

1. The limited literature on estimation of demand systems from household budget data includes Tsujimura and Sato (1964) on Japan, Bhattacharya (1967), Joseph (1968), and Ray (1980, 1982) on India, Muellbauer (1977), and Pollak and Wales (1978) on Britain.

2. As pointed out by Barten (1977), "No clear-cut decision exists about the empirical superiority of any particular demand system. The choice depends on the preferences, the aims, and the (vested) interests of the researchers."

3. As discussed in Chapter 1, the classical restrictions of

demand theory include the Slutsky symmetry conditions, the Engel aggregation condition, the Cournot aggregation conditions, the homogeneity or Euler aggregation conditions, and the negativity condition.

4. An exact-size random sample algorithm was used to accomplish this task.

5. The Bard Program employs the Davidon-Fletcher-Powell (DFP) method (Fletcher and Powell 1963) of maximization of the likelihood function. The DFP method, a steepest descent method, involves the function $(-n/2) \log|(1/n)\widetilde{\Omega}^{ri}|$. The maximization of the likelihood function is equivalent to the minimization of L. Given initial parameter guesses, the DFP method uses an iterative process to find the minimum of L. Attempts were not made to check on whether or not relative minima had been reached by varying the starting values due to the costliness of the procedure. Several practitioners present proofs of a number of theorems to show that the algorithm always converges (Fletcher and Powell 1963; Berndt et al. 1974). As with the general case of descent methods, this procedure has the property of stability, since the value of L decreases with each iteration.

6. To conform to space limitations due to the extremely large number of parameter estimates and their associated standard errors for each sample and region, it is not possible to report all statistical results here. However, in recognition of the fact that concrete evidence regarding the statistical significance of the parameters is crucial, they are available from the authors upon request.

REFERENCES

Bard, Y. 1967. Nonlinear parameter estimation and programming. IBM Contrib. Program Libr. 360D-13.6-003.
Barten, A. P. 1977. The systems of consumer demand functions approach: A review. Econometrica 45:23-51.
Berndt, E. R., B. H. Hall, R. C. Hall, and J. A. Hausman. 1974. Estimation and inference in nonlinear structural models. An. Econ. Soc. Meas. 3/4:653-65.
Berndt, E. R., and N. E. Savin. 1975. Estimation and hypothesis testing in singular equation systems with autoregressive disturbances. Econometrica 43:937-57.
Bhattacharya, N. 1967. An application of the linear expenditure system. Econ. Polit. Wkly. 2093-98.
Brown, M., and D. Heien. 1972. The S-branch utility tree: A generalization of the linear expenditure system. Econometrica 40:737-47.
Buse, R. C., and L. E. Salathe. 1978. Adult equivalent scales: An alternative approach. Am. J. Agric. Econ. 60:460-68.
Court, R. H. 1967. Utility maximization and the demand for New Zealand meats. Econometrica 35:424-46.
Deaton, A. S. 1974. A reconsideration of the empirical implica-

tions of additive preferences. Econ. J. 84:338-48.

_____. 1978. A simple non-additive model of demand. Dep.
    Appl. Econ., Cambridge Univ. Mimeo.

Fletcher, R., and M. J. D. Powell. 1963. A rapidly convergent
    descent method for minimization. Compu. J. 6:163-68.

Green, R., Z. A. Hassan, and S. R. Johnson. 1978. Alternative
    estimates of static and dynamic systems for Canada. Am. J.
    Agric. Econ. 60:93-108.

Houthakker, H. S., and L. D. Taylor. 1970. Consumer demand in
    the United States 1929-1970. Cambridge, MA: Harvard Univ.
    Press.

Joseph, P. 1968. Application of the linear expenditure system
    to N.S.S. data: Some further results. Econ. and Polit.
    Wkly. 3:609-15.

Kuh, E. 1959. The validity of cross-sectionally estimated be-
    havior equations in time series applications. Econometrica
    27:197-211.

Lau, L. J., W. L. Lin, and P. A. Yotopoulos. 1978. The linear
    logarithmic expenditure system: An application to consump-
    tion-leisure choice. Econometrica 46:843-68.

Muellbauer, J. 1977. Testing the Barten model of household con-
    sumption effects and the cost of children. Econ. J. 87:
    460-87.

Parks, R. W., and A. P. Barten. 1973. A cross-country compari-
    son of the effects of prices, income, and population
    composition on consumption patterns. Econ. J. 83:834-52.

Pollak, R. A., and T. J. Wales. 1978. Estimation of complete
    demand systems from household budget data: The linear and
    quadratic expenditure systems. Am. Econ. Rev. 68:348-59.

Price, D. W. 1970. Unit equivalent scales for specific food
    commodities. Am. J. Agric. Econ. 53:224-33.

Prochaska, F., and R. A. Schrimper. 1973. Opportunity cost of
    time and other socioeconomic effects on away-from-home food
    consumption. Am. J. Agric. Econ. 55:595-603.

Ray, R. 1980. Analysis of a time series of household expendi-
    tures surveys for India. Rev. Econ. Stat. 62:595-602.

_____. 1982. The testing and estimation of complete demand
    systems on household budget surveys: An application of
    AIDS. Eur. Econ. Rev. 17:349-69.

Redman, B. J. 1980. The impact of women's time allocation in
    expenditure for meats away from home and prepared foods.
    Am. J. Agric. Econ. 63:234-37.

Smallwood, D., and J. Blaylock. 1981. Impact of household size
    and volume on food spending patterns. USDA, ESS, Tech.
    Bull. No. 1650.

Theil, H., and R. H. Mnookin. 1966. The information value of
    demand equations and predictions. J. Polit. Econ. 74:34-45.

Tsujimura, K., and T. Sato. 1964. Irreversibility of consumer
    behavior in terms of numerical preference fields. Rev.
    Econ. Stat. 46:305-19.

CHAPTER 7

Analysis of Food and Other Expenditures Using a Linear Logit Model

A NUMBER OF ALTERNATIVE SPECIFICATIONS are available to the economic analyst wishing to estimate a complete system, but it is seldom possible to choose among them by purely objective reasoning. Differences among alternative systems correspond to different types of behavioral characteristics assumed about the representative consumer. These characteristics can often be equated to restrictions on the parameters of a system such as those implied by classical conditions. Although some specific formulations such as the translog and AIDS models allow testing of these restrictions, generally, the analyst must choose the restrictions to be imposed or tested (and the model or models to be estimated) before the analysis begins.

In Chapter 8, Eastwood and Sun present several criteria for model selection for practical policy analysis. They suggest that a model should meet the objectives of a specific problem and be the least restrictive possible given data, time, and cost considerations. Applying these criteria to the choice of a particular complete system model would lead to an evaluation of several alternative systems with respect to computational expense as well as to the relevance of their theoretical bases to both the objective of a specific problem and the consumer behavior being observed in sample data. The ideal model would therefore need to be flexible, easily constrained, and easily estimated.

Another view of the choice between alternative demand systems is presented in Chapter 1. There, Capps and Havlicek describe two ways to formulate a complete demand system: (1) specify a particular direct or indirect utility function or (2) specify the functional form of the demand equations directly and impose the classical and modern theoretical restrictions. While most complete demand systems described in Part 2 are formulated in

---

Authors of this chapter are T. J. Tyrrell and T. D. Mount.

the first way, this chapter describes a system formulated in the second way.

The linear logit specification for a demand system has the advantage of automatically satisfying the budget constraint and therefore globally satisfying Engel and Cournot aggregation. Otherwise, this model does not restrict parameters. It exhibits the flexibility desirable for practical policy analyses, but it can also be constrained to globally satisfy homogeneity and to locally satisfy symmetry. In addition, the parameters of the model can be estimated by readily available computer programs.

The description of the general model here is followed by the application of a specific form of the model to the analysis of the effects of household size and composition on food and other expenditures. Last, the estimated model is employed for forecasting changes in household budgets for food at home and food away from home as influenced by changing family size and age distributions over the next two decades.

THE MODEL

A linear logit model for the household budget is given by

$$w_i = \frac{\exp f_i(y, p_1, \ldots, p_N, Z_1, \ldots, Z_R)}{\sum\limits_{j=1}^{N} \exp f_j(y, p_1, \ldots, p_N, Z_1, \ldots, Z_R)} \qquad i = 1, \ldots, N \tag{7.1}$$

where w_i is the budget share allocated to the ith good (i.e., $P_i q_i / y$), y is income (total expenditures on all n categories of goods), p_i is the price of the ith good, Z_r is the rth household characteristic, $f_i()$ is general notation for a function that is linear in unknown parameters, and exp is the exponential transformation.

Since the sum of these n budget shares is exactly unity, the model automatically satisfies the budget constraint for all values of all explanatory variables y, p_i (i = 1, ..., n) and Z_r (r = 1, ..., R).

A specification of the model which has been most thoroughly explored for empirical applications is based on the assumption that the unknown functions are log-linear (Tyrrell and Mount 1982)

$$f_i = a_i + b_i \ln y^* + \sum_{j=1}^{n} c_{ij} \ln p_j^* + d_i \ln S_i \tag{7.2}$$

where y^* is deflated income (total expenditures), p_j^* is the deflated price of the jth good, and S_i is the equivalent household size variable defined below.

Homogeneity of degree zero is imposed in this specification by deflating income and prices. Although elasticities always take the form of a weighted sum of terms in this model, the income (E_{im}), price (E_{ii}), and cross-price (E_{ik}) elasticities implied by the log-linear specification have a particularly simple form

$$E_{im} = b_i - \sum_{j=1}^{n} w_j b_j + 1$$

$$E_{ii} = c_{ii} - \sum_{j=1}^{n} w_j c_{ji} - 1$$

$$E_{ik} = c_{ik} - \sum_{j=1}^{n} w_j c_{jk} \qquad (7.3)$$

The difference between b_i and the weighted sum of all the b_j is equal to the amount by which the income elasticity differs from unity, and the difference between c_{ii} and the weighted sum of the c_{ij} is exactly the difference between the own-price elasticity and -1. For each other price coefficient the distance to the weighted sum of similar terms equals the deviation of the cross-price elasticity from zero. Also, each of the elasticities changes as the weights (w_j) change, and the weights change in response to the values of all the explanatory variables.

The equivalent household size variable has been specified as the product of household size (S) and a household composition variable (θ_i) for each good: $S_i = S \theta_i$. The household composition term has been formulated as the exponentiated sum of cubic polynomials of the logarithms of the ages of household members:

$$\theta_i = \exp(\sum_{s=1}^{S} g_{is})$$

where exp is the exponential transformation, and

$$g_{is} = \alpha_{i0} + \alpha_{i1} \ln(age_s) + \alpha_{i2} \ln(age_s)^2 + \alpha_{i3} \ln(age_s)^3$$

This specification is attractive because after substitution, (7.2) is linear in parameters, and the characterization of the effects of household composition on budget shares is accomplished using a relatively small number of parameters. Because the equivalent household-size variable S_i was specified as the product of household size and a household composition variable, the marginal effects of household members vary over size. The use of a sum of unit consumer scales (described in Chapter 10) for an equivalent household--size variable would imply that marginal effects of household members are constant. Forsyth (1960) has suggested that this formulation is too restrictive.

The cubic polynomial in household members was designed to capture a continuous change in tastes over an average lifetime and the logarithmic transformation was made to dampen the variations in the polynomial for old ages. A number of variations of this specification have been tested including the adjustment of age by a positive constant to avoid ln (0) for data which includes newborn infants, the use of a dummy variable to capture the effects of a working spouse, and the use of distinct age polynomials for males and females (Tyrrell and Mount 1982).

Another advantage of this particular specification is that the effects of household size and age on demand can be examined separately in terms of demand elasticities (E_{is} and E_{ia})

$$E_{is} = d_i - \sum_{j=1}^{n} w_j d_j$$

$$E_{ia} = \sum_{s=1}^{S} (A_{is} - \sum_{j=1}^{n} w_j A_{js})$$

where $A_{is} = d_i [\alpha_{i1} + 2\alpha_{i2} \ln(age_s) + 3\alpha_{i3} \ln(age_s)]$.[2] An indicator of economies of scale in consumption is provided by the sum of income and size elasticities. This sum is an exact measure of the degree of economies to scale when it equals unity, but it only serves as an indicator away from unity. That is, when ($E_{im} + E_{is}$) is greater than one, equal to one, and less than one, the ith good exhibits diseconomies of scale, constant returns to scale, and economies of scale, respectively.

EFFECTS OF HOUSEHOLD SIZE AND COMPOSITION

A complete system which is derived from the constrained maximization of utility is constructed in a way that focuses on prices and income as the causes for substitution between goods. For this type of system to be applicable to a certain set of data, it must be assumed that the consumers being observed have the same utility function--or at least that any differences between utility functions are outweighed by the influence of prices and income in the explanation of expenditure patterns. In applications where these assumptions are difficult to make, utility functions can be explicitly adjusted for the influence of noneconomic factors such as household size and composition. However, even for this "adjusted" system, the focus usually remains on prices and income as the central causes for substitution between goods.

The main advantage of the direct specification approach to

demand systems is the flexibility provided for characterizing a wide range of consumer behavior. This flexibility can be put to its best use in the analysis of these noneconomic factors for which no theoretical models are yet available. In an analysis of cross-sectional data such as from consumer expenditure surveys, it is quite likely that there are more differences between household utility functions than between observed prices or incomes. Thus, the direct specification approach can provide a very appropriate model for analyzing the influence of household composition using this type of data.

Among systems which are directly specified, the linear logit is one for which the framework for dealing with interdependencies between goods is the budget constraint. Just as for theoretically derived systems this constraint is everywhere binding. However, within this framework, the linear logit model makes no restrictions on the substitution among goods.

One specification of this model was applied to expenditures by 456 households in the northeastern United States from the 1972–73 Bureau of Labor Statistics' Consumer Expenditure Survey. The exact model used was a special case of (7.2). Since cross-sectional data were used, price terms were omitted and total expenditures were not deflated. The constant 1.75 was added to all ages to restrict the logarithmic transformation of age to positive values beginning at conception and a Lagrangian interpolation technique was used to estimate and constrain the cubic age functions (Tyrrell 1983). Expenditures were grouped into six categories: food at home, food away from home, housing, clothing, transportation, and other goods and services. The model was estimated by a maximum likelihood procedure and results are given in Table 7.1.[1] Note that only five equations are fitted and presented in Table 7.1. This is because the sixth category, other goods and services, is used as the base in the estimation of the other five expenditure categories and its parameters are normalized to be equal to zero.

One difficulty with using the linear logit model is that estimated coefficients are not easily interpreted. Structural parameters of (7.2) are related to the estimates of Table 7.1 through differences and nonlinearities. Converting these estimates to elasticities implies weighting parameters by predicted budget shares which depend on specified values of explanatory variables. Computed elasticities are given in Table 7.2 for food at home and food away from home for three types of households with the approximate family size composition and total budget of households in the sample with heads of the household aged 20, 35, and 60.

These results indicate that the budget share going to food at home increases over age, changing from 14 percent to 19 percent between ages 20 and 60 for couples with virtually the same total budget. Household composition and income effects are illustrated by the family of four with a significantly larger

TABLE 7.1.   Linear logit budget allocation model parameter estimates

Coefficient[a]	Food at home	Food away	Housing	Clothing	Transportation
$a_i - a_6$	4.310	-4.250	0.780	-1.894	-1.964
	(2.419)[b]	(4.227)	(2.400)	(3.414)	(2.630)
$b_i - b_6$	-0.611	0.334	-0.0985	0.0749	0.165
	(0.288)	(0.488)	(0.281)	(0.397)	(0.305)
$d_i - d_6$	0.515	-0.305	-0.297	0.0828	0.0645
	(0.371)	(0.585)	(0.354)	(0.483)	(0.370)
$d_i g_{im0} - d_6 g_{6m0}$	-1.123	-0.434	0.192	-0.193	0.841
	(3.158)	(4.929)	(2.712)	(3.875)	(2.901)
$d_i g_{im14} - d_6 g_{6m14}$	0.0377	0.0134	0.0766	0.0077	-0.0686
	(0.0994)	(0.171)	(0.0949)	(0.133)	(0.107)
$d_i g_{im65} - d_6 g_{6m65}$	-0.0429	-0.199	-0.711	-0.427	-0.300
	(0.881)	(1.511)	(0.847)	(1.238)	(0.951)
$d_i g_{if0} - d_6 g_{6f0}$	-0.744	0.499	-0.0469	0.189	-0.0655
	(2.194)	(3.431)	(1.967)	(2.697)	(2.187)
$d_i g_{if14} - d_6 g_{6f14}$	-0.0028	-0.0063	0.0249	-0.0109	-0.0035
	(0.0967)	(0.165)	(0.0935)	(0.127)	(0.0994)
$d_i g_{if65} - d_6 g_{6f65}$	0.375	-1.468	0.0171	-0.703	-1.047
	(0.801)	(1.500)	(0.779)	(1.183)	(0.930)

Note: Locations inside SMSAs of more than one million people in the
Northeast but outside central cities; 456 sample consumer units, 2,235 degrees
of freedom.
[a]Parameters of equation (7.2) expressed as differences from those of the
other goods and services equation (j = 6). Corresponding to the values of the
age polynomials at ages 0, 14, and 65 are g_0, g_{14}, and g_{65}. Coefficients
for age 20 were constrained to zero.
[b]Asymptotic standard errors in parentheses.

TABLE 7.2.   Elasticities for food at home and food away from home for three
             typical households in the Northeast in 1972

	Young couple[a]		Family of four[b]		Retired couple[c]	
	FAH[e]	FAFH[f]	FAH	FAFH	FAH	FAFH
Elasticity						
Income[d]	.44	1.39	.48	1.22	.48	1.43
Size	.51	-0.31	.46	-0.36	.47	-0.36
Age	.00	-0.01	-.04	0.00	.04	-0.07
Budget share	.14	0.06	.20	0.05	.19	0.04

Note: From Table 4.15 of Tyrrell (1979).
[a]One male age 20; one female age 20; total budget = $6,562.
[b]Two males ages 35 and 10; two females ages 35 and 10; total budget = $8,684.
[c]One male age 60; one female age 60; total budget = $6,730.
[d]Income elasticity is used to mean total expenditure elasticity.
[e]Food at home.
[f]Food away from home.

income which spends the largest share of the three households on
food at home. Food consumed away from home is shown to be a
luxury good with income elasticities from 1.22 to 1.32, while
food at home is shown to be a necessity with income elasticities
between .44 and .48.

Size and age elasticities are relatively similar between
household types but remarkably different between goods. A
marginal increase in family size will increase expenditures on
food at home and reduce those on food away from home. For both
expenditure categories, we can conclude that returns to scale in

consumption are approximately unitary although returns seem to be greater to food at home for couples and to food away from home for the family of four. Conclusions which might be drawn from the age elasticities are the same as indicated by predicted budget shares. Expenditures on food at home generally increase with age while expenditures on food away from home decrease.

FORECASTING HOUSEHOLD FOOD BUDGETS

Since 1960, U.S. per capita food expenditures have more than tripled, increasing from $411 to $1,337 in 20 years. However, percentage allocations of expenditures on food consumed at home have declined from 17 percent to 11 percent of total expenditures, while food consumed away from home have remained relatively stable at 6-7 percent (Table 7.3). The linear logit model has been applied to the problem of predicting the effects of future changes in household composition on food consumption expenditures.

TABLE 7.3. Percentage of total U.S. personal consumption expenditures by category and year

Category	1960	1970	1980
Clothing	10	9	8
Transportation	13	13	14
Utilities	12	12	12
Housing	15	15	16
Food at home	17	13	11
Food away from home	6	6	7
Other goods and services	27	32	32
Total	100	100	100

Source: U.S. Bureau of the Census (1962, 1972, 1982).

Among the potential causes for the budget shift away from food at home are increases in consumer income and changes in population characteristics such as household size and age distributions. It is reasonable to believe that, taken together, these factors could explain a substantial portion of the historical shifts in expenditures.

For forecasting future expenditures, the difficulty is to sort out the separate influences of causal factors so that behavioral responses to probable future scenarios can be predicted. As discussed earlier, the complete systems approach is particularly useful for the "sorting-out" process and, depending on the hypothesized behavior and the factors being investigated, the forecaster has a wide variety of complete systems models to choose from.

For example, if the effects over time of income on food budgets have the relative magnitudes as estimated by Craven and Haidacher in Chapter 4 (Table 4.4), which are very similar to those estimated between households (Table 7.2), then the shift away from food consumed at home during this period has been caused, at least in part, by the increase in income and a shift toward luxury goods.[2]

While changes in relative prices between goods also have the potential for influencing budget shifts, it does not appear that these have been a major cause for the shift in food expenditures. On one hand, casual inspection of national average price data suggests that the relative prices for the two food categories have changed almost proportionately so that we would not expect a large direct substitution effect. On the other hand, while prices for goods such as fuel and utilities have recently experienced dramatically larger increases, cross-price elasticity estimates presented in Chapter 4 suggest that indirect substitution effects may be relatively small.[3] Judging by this evidence, it seems likely that the shift away from food at home (Table 7.3) has been influenced more by income than price effects.

The two other factors which were mentioned as potential causes for the shift in the food budget are changes in household size and the age distribution. The estimates from the previous section provide some measure of patterns over households. However, for such estimates to be used for forecasting, it must be assumed that these estimated differences do not change over time.

The effects which would have been predicted for the changes observed in data for 1960 and 1978 are roughly consistent with the expenditure patterns observed in Table 7.3. That is, the increase in the percentage of one- and two-person households from 11.9 percent to 18.4 percent and the movement of the post-war baby boom generation through the lower age categories would each have been expected to shift expenditures away from food consumed at home.

Under the assumption that the estimated linear logit model describes future patterns of household expenditures, the _ceteris paribus_ effects of changes in household size and age distributions have been forecasted. This analysis was conducted by simulating the process of sampling a population undergoing these changes, where characteristics were assigned to individual households according to their relative frequencies in the population. The estimated model was then used to predict expenditures by these households.

The initial population characteristics were taken from the 456 households observed in 1972 and used in the estimation of the model. The relative frequencies of alternative values for nine household characteristics were computed. These included frequency distributions for (1) the age of the head of a household, (2) the sex of the head given age, (3) the income of the household given the age of its head, (4) the sex of the head given age, (5) the household size given the age of its head, (6) the age of the second member given the age of the head, (7) the sex of the second member given the difference between age and the age of the head, (8) the age of other members given the age

of the head, and (9) the sex of the other members given their
ages.

For each year beginning in 1972, characteristics for 456
households were generated by applying uniformly distributed (0,
1) random numbers to the frequencies and joint frequencies. To
simulate gradual changes in the characteristics of the population being sampled, the age and household-size frequencies were
changed according to intermediate level projections from
the U.S. Bureau of Labor Statistics.[4]

The average results for food expenditures at home and away
from home for 1972 to 2000 are shown in Figure 7.1. The major
conclusion is that with constant prices and a fixed distribution
of income over the ages of household heads, the effect of the
projected decline in household size and upward shift in the age
distribution will be to reduce the average household's percentage allocation to all food (from 23.6 percent to 21.3 percent)
even though the percentage to food away from home increases
(from 5.1 percent to 5.9 percent).

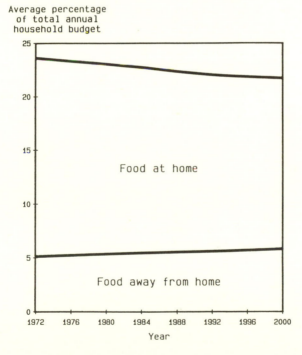

Fig. 7.1. Household food budget projections: 1972-2000.
Projections are average values from a simulated sample of 456
households in the Northeast with changes in household size and
age distribution derived from Census Bureau Household Series B
and Population Series II.

While the overall effect on food expenditures shown in Figure 7.1 can be explained in part by the fact that 456 households contain fewer total persons each year, it is clear that household size and age composition are likely to have a substantial effect on allocations between food categories during the rest of this century.

CONCLUSIONS

Chapter 1 describes two ways of formulating a complete demand system: by solving the constrained maximization problem for a specified utility function or by direct, but arbitrary, specification of a system of demand equations. The linear logit model for household expenditures falls somewhere between the two. Although it does not always satisfy the maximum conditions for a specific utility function it is everywhere constrained by the budget.

One form of this model was applied to data from the 1972-73 BLS Consumer Expenditure Survey to explore the joint effects of household income, size, and age characteristics on food budgets. These cross-sectioned results are consistent with the notion that the budget shift away from food consumed at home observed over the past two decades has been due to a combination of the effects of increasing income and declining household size. Further, this trend is projected to continue to the year 2000 if prices are stable or increase in proportion to one another, even if real income does not continue to grow. This result is due primarily to the decline in household size projected through the rest of the century.

NOTES

1. The likelihood function was based on the assumption that budget shares are allocated according to a multinomial distribution. For details, see Tyrrell (1979). Less restrictive assumptions were employed in Tyrrell and Mount (1982).

2. After accounting for general price increases, per capita real income increased by 64 percent between 1960 and 1978, from $2,517 to $4,131 in 1972 dollars.

3. Coughlin (1982) explored direct and indirect price effects on food using the multinomial logit model.

4. In U.S. Bureau of the Census (1979) based on household series B, and population series II, the average household size, between 1975 and 1995, is projected to decline from 2.94 to 2.39 and the average age of the population over 15 to rise from 40 to 43.

REFERENCES

Coughlin, R. 1982. An analysis of how higher fuel prices affect
    the cost of living for different types of families in the
    northeastern U.S. Thesis, Cornell Univ.
Forsyth, F. E. 1960. The relationship between family size and
    family expenditure. J. R. Stat. Soc. Series A, 123:367-97.
Office of the President. 1979. Economic report of the Presi-
    dent. Washington, DC: U.S. Government Printing Office.
Tyrrell, T. J. 1979. An application of the multinomial logit
    model to predicting the pattern of food and other household
    expenditures in the Northeastern United States.   Ph.D.
    diss., Cornell Univ.
_____. 1983. The use of polynomials to shift coefficients in
    linear regression models. J. Bus. & Econ. Stat. 1:249-52.
_____, and T. D. Mount. 1982. A nonlinear expenditure system
    using a linear logit specification. Am. J. Agric. Econ.
    64:539-46.
U.S. Bureau of Labor Statistics. 1977.   1972-73 Consumer expen-
    diture survey: Interview survey summary public use tape.
    Washington, DC: U.S. Government Printing Office.
U.S. Bureau of the Census. 1979. Projections of the number of
    households and families: 1979 to 1995. Current population
    reports. Ser. P-25. No. 805. Washington, DC: U.S.
    Government Printing Office.
_____. Statistical abstract of the United States. 1962, 1972,
    and 1982 eds. Washington, DC: U.S. Government Printing
    Office.

CHAPTER 8

Complete Demand Systems and Policy Analysis

ADVANCES IN APPLIED DEMAND ANALYSIS are enabling economists to narrow the gap between the theory of consumer behavior and the application of statistical tools to available data to generate estimates of consumer demand parameters. Among these developments are efforts by economists to incorporate the inherent simultaneity of consumer purchase decisions across the spectrum of goods and services into the estimation process. Models estimated within this context comprise an improved framework for the empirical analyses of problems dealing with the economic interrelatedness inherent in consumer demand. Such analyses include program and policy issues which entail manipulating determinants of consumer purchases.

Two general types of frequently encountered problems are well suited to be analyzed within a complete demand systems context. One pertains to programs or policies that use prices as the instruments for change, and the other uses quantities as the instruments. In the first case, prices are manipulated or administered, and these changes have impacts on quantities bought or consumed. In the second case, quantities are purchased or otherwise altered in an attempt to influence market prices.

A common feature of both types of problem is that the variables which are controlled comprise a subset of the prices and quantities which ultimately determines demand. However, due to the inherent simultaneity of consumer decision making, all goods and services consumed are affected. The nature and magnitude of these effects involve an empirical question which should be

Authors of this chapter are David B. Eastwood and Theresa Y. Sun. Eastwood was on leave with the Food Demand Research Section, USDA, when this research was conducted.

considered in the development of policies and the operation of programs. Consequently, a complete demand systems framework is warranted.

The process of using a complete demand system to evaluate policy issues can be separated into four parts. One is the choice of the appropriate complete demand system to be used. The second is adaptation of the estimated demand model, to permit development of an empirical framework so the policy issue can be addressed. Third is use of an elasticity matrix to answer problems from a quantity dependent perspective. Fourth is use of a flexibility matrix to answer issues from a price dependent perspective.

This chapter focuses attention on the second part of the process. A methodology is outlined adapting an estimated set of response parameters, obtained from estimating a demand system, to provide an improved means for evaluating proposed policies and program operation. The problem can be envisioned as one of matching the categories of the estimated system to those involved in the policy analysis. Theoretical and empirical restrictions dictate the demand system categories while policy variables are determined by a different set of forces. Consequently, a mechanism is needed to bridge the two.

The starting point is the model presented in Chapter 1, stated here in elasticity form. This reference point permits a clear distinction between the use of an elasticity matrix to answer quantity dependent policy questions and the use of a flexibility matrix to answer price dependent policy questions. Then the discussion turns to a procedure for bridging the gap between the demand and policy categories. The last section provides an illustration of the procedure. Specifically, the organization of the chapter is as follows: (1) demand systems, (2) elasticities and analysis of price changes, (3) flexibilities and analysis of quantity changes, (4) reconciling demand and policy categories, (5) an example, and (6) summary.

DEMAND SYSTEMS

The solution of the consumer's utility maximization problem is shown in Chapter 1 to be a system of demand equations. This system also can be expressed in percentage terms as

$$\%\Delta q_1 = e_{11}(\%\Delta p_1) + \ldots + e_{1n}(\%\Delta p_n) + \eta_1(\%\Delta y)$$
$$\vdots$$
$$\%\Delta q_n = e_{n1}(\%\Delta p_1) + \ldots + e_{nn}(\%\Delta p_n) + \eta_n(\%\Delta y) \tag{8.1}$$

or in matrix notation

$$\%\Delta q = E(\%\Delta p) + \eta(\%\Delta y) \tag{8.2}$$

As in Chapter 1 the e_{ij} are uncompensated price elasticities
and η_i is the ith income elasticity. The general restrictions of
homogeneity, symmetry, and adding-up are assumed to hold. In
order to reduce the number of parameters which have to be estimated and to accommodate data limitations, additional assumptions are imposed to estimate various models. Other factors
which also serve to reduce the number of parameters are time and
costs associated with data collection and estimation procedures
because these two factors vary in the same direction as the
number of goods included in the demand model.

Throughout the remainder of this chapter, it is assumed that
a demand system has been estimated. The problem discussed is
one of making adjustments in the estimated model to permit the
analysis of policy issues which are defined for similar but not
identical goods. An effort is made to provide a procedure which
preserves the general theoretical properties of the demand model
as much as possible.

ELASTICITIES AND ANALYSIS OF PRICE CHANGES

A clear result of the quantity dependent model is that the
price of any particular good has the potential to affect to
varying degrees the quantities demanded of every good purchased
by consumers. Consequently, policy analysis entailing the
manipulation of prices and/or income to produce changes in the
quantities demanded should attempt to reflect the simultaneity.
The advantage of using a complete demand system framework is
that these effects can be traced across all demand categories
via E and η. A column of E represents the impact of a percentage change in the price of a good <u>ceteris</u> <u>paribus</u> on the
various quantities, q_i, ..., q_n. The effect of a percentage
change in p_1 on the percentage change in each of the quantities
is

$$\%\Delta q_i = e_{i1}(\%\Delta p_1) \quad \text{for} \quad i = 1, \ldots, n \qquad (8.3)$$

A row of E represents the combined effects of percentage changes
in the prices of each of the n goods on the quantity demanded of
the good associated with that row. Assuming no change in income

$$\%\Delta q_i = \sum_{j=1}^{n} e_{ij}(\%\Delta p_j) \qquad (8.4)$$

The percentage changes can be converted easily to changes in
the levels of the quantities demanded via multiplication of the
simulated percentage changes in quantity by the respective quantity levels. Appropriate quantity levels are determined by the
data used to generate the elasticity estimates.

FLEXIBILITIES AND ANALYSIS OF QUANTITY CHANGES

Complete demand systems can also be used for policy analyses in which quantities are manipulated and interest focuses on the resultant price changes. Such analyses have the reversed causality of that represented by (8.2). This reversed causality is represented by (8.5). The general theoretical properties of price dependent demand systems have been discussed most recently by Anderson (1980) who outlines the constraints corresponding to homogeneity, symmetry, and adding-up. Huang (1983) has derived inverse demand systems for a set of well-known utility functions.

$$\%\Delta p = F(\%\Delta q) + \delta(\%\Delta y)$$

$$F = E^{-1}$$

$$\delta = -E^{-1}\eta \qquad\qquad\qquad\qquad (8.5)$$

The matrix F is usually called the flexibility matrix. Traditionally, it has been obtained by inverting E.[1] However, the recent works of Anderson and Huang have provided the framework for estimating F directly. Regardless of which approach is taken, the interpretation of F is analogous to that outlined for E. A column of F represents the effects of a percentage change in the quantity of one good on each price. Should there be, for example, a percentage change in the third quantity category, the effect on each of the prices is

$$\%\Delta p_i = f_{i3}(\%\Delta q_3) \quad \text{for} \quad i = 1, \ldots, n \qquad (8.6)$$

A row of F represents the effects of percentage changes in all of the n quantities on a specific price. The sum of each of the percentage quantity changes on the percentage change in the ith price is

$$\%\Delta p_i = \sum_{j=1}^{n} f_{ij}(\%\Delta q_j) \qquad\qquad\qquad (8.7)$$

Thus, F allows decision makers to evaluate policies and programs which directly control quantities and have impacts on prices. Conversion from percentage changes to changes and levels follows the procedure discussed above for the quantity dependent simulation.

RECONCILING DEMAND AND POLICY CATEGORIES

Viewed from a policy perspective, the frequently encountered problem is one of tailoring an estimated demand system so it can be used to analyze the effects of policy controlled variables. The categories of goods of the demand system usually do not correspond exactly to those which are the instruments of change. Among the causes of the discrepancies is the problem that available data, especially time-series data, may not provide a fine enough breakdown of retail goods and services to estimate the specific elasticities or flexibilities associated with the policy. Furthermore, some demand system models from which elasticities or flexibilities are derived (e.g., the linear expenditure system) are inherently restricted to broad aggregates.

The process of using complete demand systems in policy analyses begins with selection of an appropriate complete demand system. Choice of a system requires that one be familiar with theoretical properties of the various models so the analyst can choose that complete demand system for which the properties are most compatible with the issue to be evaluated. Implicit in this approach is the assumption that either the appropriate model has been estimated or there is enough time to estimate the chosen model to obtain the corresponding response parameter estimates. Which model to use varies from issue to issue depending upon the biases of the analyst and the detail required. Broad policy questions often can be simulated by the more aggregate models, whereas, disaggregate systems typically are used for more specific issues.

Present interest focuses on a methodology for matching the categories associated with the policy problem to those categories of goods included in the estimated system. At first blush, one might be inclined to insert rows and columns as subcategories in E or F for the needed policy variables. However, this is not recommended for two basic reasons. The first is that the data which are employed to generate the additional elasticities or flexibilities most likely are not consistent with the data used to estimate the demand system. Otherwise, the subcategory could have been broken out as a specific item when estimating the demand system. The second reason is that there is no unique way to insert these additional values into the matrix so as to preserve the theoretical properties of the model. The general constraints of homogeneity, symmetry, and adding-up, or their flexibility counterparts, define sets of relations among the elasticities or flexibilities. There is no unique way of inserting a row or column into an estimated elasticity or flexibility matrix so as to preserve these relationships. Through disrupting the constraints, such an endeavor would detract from the purpose of using complete demand systems in policy analyses.

The methodology described below refers to a common situation involving three factors: (1) quantities comprise the policy instruments, (2) a quantity dependent demand system was estimated to generate E, and (3) E was inverted to obtain F.

The breakdown of the policy instruments comprised a detailed set of food categories.[2] In fact, the detail was so fine that time-series data on price and quantity for many of the policy goods were not available at the retail level. A further complication was that the available policy data pertained to prices

and quantities which were actually bought at the wholesale level
and at irregular intervals.

Clearly, this precluded using these categories in the
estimation of a complete demand system. It further dictated the
way in which the detailed policy instrument quantities related
to the chosen demand system aggregates. The paucity of price
data necessitated considering the quantity demanded of the ith
good in the demand system to be a weighted average of its
various subcategories. The weights are the proportions of the
subcategories in the respective totals (8.8).

$$q_i = \sum_{k=1}^{K_i} w_{ik}q_{ik}$$

$$w_{ik} = q_{ik}/q_i \tag{8.8}$$

where K_i = the number of subgroups in commodity i. When dealing
with percentage changes in the quantity demanded, this expression
becomes

$$\%\Delta q_i = \sum_{k=1}^{K_i} w_{ik}(\%\Delta q_{ik}) \tag{8.9}$$

Incorporating (8.9) into the system of flexibility relationships, one can then estimate the effect of a change in a quantity
instrument variable across the spectrum of retail demand. The
procedure (8.10) is to compute the change in the subcategory as
a percentage of the respective aggregate included in the model,
and then, to multiply the appropriate weighted percentage change
and flexibility expressions.

$$\%\Delta p_i = f_{ij}[w_{jk}(\%\Delta q_{jk})] \tag{8.10}$$

This methodology constitutes the bridge for adapting estimated complete demand systems to analyses of policy problems.
Schematically, the process is shown as Figure 8.1. The weighting procedure is seen as the transformation of the instrument
quantity changes into the respective demand model aggregate so
the response parameters (in the present case flexibilities) can

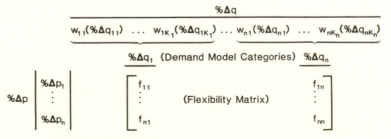

Fig. 8.1. The weighting procedure.

be used to assess impacts across demand categories. One advantage of this approach is that the simultaneity inherent in
consumer decision making is incorporated into the analyses.
Another is the theoretical properties of the estimated system
are preserved. These advantages suggest an enhanced policy
analysis capability through adapting complete demand systems.

AN EXAMPLE

The U.S. Department of Agriculture in compliance with Section
32 of the Agricultural Adjustment Act has authority to purchase
food commodities deemed to be in surplus. Food purchased under
this program is distributed to public nutrition programs. These
purchases have the effect of removing commodities which otherwise
would become available to consumers. The policy intervention
variables of this ongoing program are quantities which in turn
affect prices. Because of the simultaniety in food demand, the
impact of any quantity removal should be traced across food
purchases.

A computer simulation program known as the Removal Impact
Model (RIM) is designed to evaluate the effects of alternative
purchase strategies.[3] Implementation of Section 32 purchases
are evaluated in a two-stage process. The first-stage deals
with minimum purchases which represent the minimum quantities
required for the operation of food assistance programs regardless
of market conditions. The second stage reflects additional
purchases designed to enhance prices or farm income through
market-oriented quantity-price adjustments. Present interest
centers on illustrating the weighting methodology so the discussion is limited to stage one purchases. Since the policy instruments are quantities and their impacts on prices are to be
simulated, a price dependent (flexibility) model is needed.

The nature of Section 32 purchases dictates that a disaggregate complete demand system model be used due to the detail
associated with the purchases. The specific model chosen has
been developed by Huang (1980), and there are several
considerations involved with this choice.[4] It is a very disaggregate system, having 42 food commodity categories and one
nonfood category. The specific goods provided the closest approximation to the categories associated with purchases under
the program. Second, even though it is a very disaggregate
model, the general theoretical constraints of homogeneity,
symmetry, and adding-up are imposed as part of the constrained
maximum likelihood algorithm. Third, this estimated disaggregate system utilizes one set of data to obtain values for all
the elasticities. Other disaggregate models such as Brandow
(1961) and George and King (1971) have utilized several dis-
parate data sources.

Because the program evaluation to be addressed resembles the

price dependent form represented by (8.5), it is assumed that
the 43 x 43 elasticity matrix could be inverted to obtain a
theoretically consistent matrix of flexibilities. Space limitations preclude presentation of the complete 43 x 43 element
matrix. Only those elements of F which are used in (8.6)-(8.10)
are reported.

The bridging problem is displayed in the first two columns
of Table 8.1. The left-hand column lists the 42 goods associated with the demand system. The next column presents the corresponding RIM categories of which there are 61. This table is
arranged in such a way as to show the matching between the
demand and policy categories. The absence of one-to-one corre-
spondence between the categories of the chosen demand system and
RIM categories shows the need for a reconciling procedure in
order to be able to simulate the effects of the quantity removals
on retail prices.

The approach begins with the computation of the w_i. Given
the nature of the problem, various data sources were used as
noted in the table. In several cases, data limitations caused
some lack of comparability between commodity definitions for the
associated demand model and RIM categories. It should also be
noted that whenever the surplus commodity categories do not
comprise the entire set of categories of the retail demand
group, the sum of these weights is less than one. For example,
the RIM purchase categories of beef do not comprise all of the
components of the aggregate of beef and veal retail demand.
Therefore, the sum of these w_{ik} is less than one.

Estimates of the percentage change in price using (8.10)
also required the calculation of percentage changes in quantity.
These were generated as the amounts purchased under the program
divided by the available supply of the associated retail demand
category.

Minimum purchases for each of the RIM commodities (q_i) are
also presented in Table 8.1. Assuming that these purchases
occur, the program of purchases results in a decrease in retail
market supply. Converting them into percentage of consumption
and weighting by the appropriate values produces the data
appearing in the weighted percent column.[5] For example, the
weighted percent of 45,000 tons of frozen ground beef is 0.037.

Given these percentage changes, one can estimate the impact
of the respective purchase on any retail price in the demand
system using (8.10). Referring to Figure 8.1, this identifies
the respective column of the flexibility matrix. To estimate
the impact on a specific retail price, one only has to multiply
by the appropriate flexibility of that column. Continuing with
the frozen ground beef example, the estimated impact on the
percentage change in the retail price of beef and veal is

TABLE 8.1. Retail demand system versus RIM commodities

Retail demand system commodities	RIM commodities	w_{ik}	Minimum q_i (tons)	Weighted percent	$\%\Delta P_i$
1. Beef and veal	1. Frozen ground beef	0.0855a	45,000	0.037	0.03
	2. Frozen beef patties	0.0855a	0		
	3. Cooked patties	0.0855a	0		
	4. Frozen beef roast	0.0386b	0		
	5. Canned roast	0.0386b	9,050	0.003	
	6. Canned roast in juice	0.0386b	0		
2. Pork	7. Frozen pork roast	0.0952c	0		0.12
	8. Canned pork	0.0952c	0		
3. Other meat	9. Frankfurters	0.2210d	0		1.60
4. Chicken	10. Cut up chicken	0.3832e	41,000	0.332	0.71
	11. Breaded chicken	0.0725f	0		
	12. Boned-canned chicken	0.0725f	12,500	0.019	
5. Turkey	13. Whole turkey	0.0925g	19,500	0.183	0.16
	14. Rolled turkey	0.0047g	12,700	0.006	
6. Fish (fresh, frozen)					
7. Fish (canned, cured)					
8. Eggs	15. Dried eggs	0.0045g	1,900	0.002	0.44
9. Milk (fluid and cream)					
10. Milk (evaporated, condensed, and dry)	16. Dry milk	0.2194g	10,500	0.138	0.64
11. Cheese	17. American processed cheese	0.3773g	28,000	0.466	5.18
	18. Mozzarella cheese	0.0161g	0		
12. Frozen dairy					
13. Butter	19. Butter	1.0000	23,850	5.049	17.63
14. Margarine	20. Margarine	1.0000	0		
15. Shortening and other fats	21. Lard	0.0037g	0		0.21
	22. Shortening	0.1031g	8,900	0.021	
	23. Peanut butter	0.1886g	10,300	0.044	
	24. Peanut oil	0.0043g	11,400	0.001	
16. Fresh apples					
17. Fresh oranges					
18. Fresh bananas					
19. Fresh grapes					
20. Fresh grapefruit					
21. Cantaloupes and watermelons					
22. Canned fruit cocktail	25. Fruit cocktail	1.0000	0		3.89
23. Canned, frozen, citrus juices	26. Canned orange juice	0.0335h	9,325	0.007	0.01
	27. Lemon juice (frozen)	0.0642h	3,456	0.005	

TABLE 8.1 (continued)

Retail demand system commodities	RIM commodities	w_{ik}	Minimum q_i (tons)	Weighted percent	$\%\Delta P_i$
24. Canned, processed other fruit	28. Canned apricots	0.0179h	0		0.13
	29. Canned pears	0.0642h	16,312	0.084	
	30. Canned peaches	0.1519h	32,712	0.163	
	31. Canned cranberries	0.0208h	5,048	0.003	
	32. Canned apples and applesauce	0.0715h	26,787	0.083	
	33. Raisins	0.0417h	0		
	34. Prunes	0.0149h	0		
	35. Canned grapefruit sections	0.0179h	0		
25. Lettuce					
26. Tomatoes					
27. Celery					
28. Onions					
29. Cabbage					
30. Fresh potatoes	36. White round potatoes	0.0001j	2,590	0.000	0.06
	37. White french fried potatoes	0.2007h	39,870	0.094	
	38. White dried potatoes	0.0163h	9,322	0.002	
	39. Canned sweet potatoes	0.0100h	9,564	0.001	
	40. Dried sweet potatoes	0.0001i	0		
31. Carrots and other vegetables					
32. Canned tomatoes	41. Canned whole tomatoes	0.2212h	13,388	0.122	0.33
	42. Tomato juice	0.1372h	7,300	0.041	
	43. Tomato catsup	0.3142h	16,196	0.210	
	44. Tomato paste	0.2876h	15,004	0.179	
33. Canned and frozen peas	45. Frozen peas	0.2804h	5,225	0.252	0.34
	46. Canned peas	0.7196h	9,600	1.187	
34. Canned corn and other processed vegetables	47. Frozen corn	0.0531h	3,284	0.004	0.00
	48. Canned corn	0.1891h	5,634	0.004	
	49. Frozen snap beans	0.0300h	9,500	0.028	
	50. Canned snap beans	0.1771h	2,250	0.007	
	51. Beets	0.0420i	0	0.010	
35. Wheat flour (white, semolina)	52. All purpose	0.4067j	75,270	0.026	2.38
	53. White flour	0.0837j	50,180	0.004	
	54. Durham flour	0.0949j	12,544	0.001	
	55. Macaroni	0.0756j	1,400	0.000	
	56. Spaghetti	0.0756j	0		
36. Rice	57. Rice	1.0000h	8,550	1.000	0.80
37. Dry beans and peas	58. Dried snap beans	0.7317h	10,000		26.50
	59. Dried peas	0.0244h	0		

TABLE 8.1 (continued)

Retail demand system commodities	RIM commodities	w_{ik}	Minimum q_i (tons)	Weighted percent	$\%\Delta P_i$
38. Corn and other cereals	60. Oats	0.2038h	4,000	0.006	0.22
39. Sugar					
40. Other sweeteners					
41. Coffee					
42. Other beverages	61. Apple juice	0.1006h	0		
43. Nonfood					

aOne-third of CPI cost weight for ground beef relative to all beef.
bOne-third of CPI cost weight for chuck roast relative to all beef.
cOne-third of CPI cost weight for other pork.
dCPI cost weight.
eProportion of CPI cost weight for chicken parts.
fProportion of nonbroiler production to total broiler production (Poultry Production Distribution and Income, 1978).
gChain Store Age (1978).
hJohnson (1978).
iUnpublished estimates for Spring 1980 were obtained from the Food and Agriculture Branch, USDA.
jRelative proportion of different types of flour to consumption of wheat and rye flour (per capita basis).

$$\%\Delta p_{bv} = f_{bv,bv}[w_{bv,g}(\Delta q_g/q_{bv})]$$

$$= -2.0767(-0.00037)$$

$$= 0.0008$$

where bv = beef and veal, g = frozen ground beef, and $f_{bv,bv}$ = -2.0767. A decrease of 0.037 percent in the quantity of beef and veal available in the marketplace due to the stage one purchase of frozen ground beef is estimated to generate a .08 percent increase in the retail price of beef and veal.

Of course, the effect on price of a percentage change in the quantity of beef and veal is not limited to its own category. Consumers react to the price change by altering their purchased bundles of food items. The other flexibilities in the beef and veal column measure these effects on the respective retail prices. For example, we can estimate the effect of the beef removal on pork prices. Let pk denote pork, and $f_{pk,bv}$ = -2.891.

$$\%\Delta p_{pk} = f_{pk,bv}[w_{bv,g}(\Delta q_g/q_{bv})]$$

$$= -2.891(-0.00037)$$

$$= 0.00011$$

A decrease of 0.037 percent in the quantity of beef and veal is estimated to generate a 0.01 percent increase in the price of pork.

The usefulness of adding up related changes can also be illustrated. That is, suppose interest is in the combined effects of several commodity purchases on a selected retail price. It can be obtained three ways, by (1) using the appropriate elements in the respective row of the flexibility matrix, (2) multiplying by the weighted percentage change in quantities, and (3) summing. This is the equivalent of applying (8.10) and (8.7). For example, stage one purchases of dried eggs (de), dried milk (dm), and canned peas (cp) have a combined impact on the retail price of eggs. Let e, m, and peas denote the retail commodities and -6.6672, -0.4532, and -0.1024 the flexibilities.

$$\%\Delta p_e = f_{e,e}[w_{e,de}(\Delta q_{de}/q_e)] + f_{e,dm}[w_{m,dm}(\Delta q_{dm}/q_m)]$$

$$+ f_{e,cp}[w_{peas,cp}(\Delta q_{cp}/q_{peas})]$$

$$= -6.6672(-0.0000227) - 0.4532(-0.00138)$$

$$-0.1024(-0.01187)$$

$$= 0.0019.$$

The combined effect of the percentage changes on the percentage change in the price of eggs is estimated to be .19 percent.

The last column of Table 8.1 presents the combined effect of
stage one purchases on each of the retail categories associated
with RIM purchases. For example, 0.03 for beef and veal suggests that all of the stage one purchases combined results in a
0.03 percent increase in the retail price of beef and veal.
Inspection of the column reveals that the largest combined
effect is on dry beans and peas (26.5 percent) and is smallest
for canned corn and other processed vegetables (0.000).

SUMMARY
    This chapter has focused on a procedure which permits the
application of a complete demand system to generate estimates of
the effects of programs and policies in retail markets. The
methodology is a way of matching the demand system categories to
those of the policy issue. The objective is to leave the demand
system as estimated so as not to alter the properties of the
model. A weighting scheme is used to transfer between the demand
categories and the intervention categories. The weighting scheme
is developed for quantity intervention analyses, but it could be
manipulated to accommodate price intervention strategies. An
illustration of the technique is provided via partial analysis
of Section 32 purchases by the U.S. Department of Agriculture.

NOTES

    1. It is assumed in this case that the computation of E^{-1}
produces an F which is appropriate for the intervention analysis.
    2. Although this is a specific situation, the methodology
is easily transferred to situations involving prices as policy
instruments.
    3. For a discussion of the original RIM, see Haidacher et
al. (1975).
    4. The theoretical derivation, estimation technique, and
data are discussed in Huang (1980).
    5. The quantity data are total consumption for each of the
42 food categories for 1978.

REFERENCES

Anderson, R. W. 1980. Some theory of inverse demand for applied
    demand analysis. Eur. Econ. Rev. 14:281-90.
Brandow, G. E.   1961.   Interrelation among demands for farm pro-
    ducts and implications for control of market supply. Pa.
    Agric. Exp. Stn. Bull. No. 680.
Crop Reporting Board. 1978 Poultry production, distribution and
    income. USDA Stat. Bull. No. 602.

Editors of Chain Store Age. 1978. Chain store age supermarkets,
     1978 sales manual. New York: Lebhar-Friedmand, Inc.
George, P. S., and G. A. King. 1971. Consumer demand for food
     commodities in the U.S. with projections for 1980. Giannini
     Found. Monogr. 26, Univ. of California, Berkeley.
Haidacher, R., R. C. Kite, and J. L. Matthews. 1975. Applica-
     tion of a planning decision model for surplus commodity
     removal programs. In Quantitative models of commodity
     markets, ed. Walter Labys. Cambridge, MA: Balinger.
Huang, K. 1980. Estimation of a complete food demand system
     with limited sample observations. Manuscr., USDA, Food
     Demand Res. Sect., Food Econ. Branch.
_____. 1983. The family of inverse demand systems. Eur. Econ.
     Rev. 23:329-37.
Johnson, A. O. 1979. Food consumption, prices and expenditures.
     USDA Agric. Econ. Rept. No. 138, 1977 Supplement.

PART III

Partial Systems: Factors Influencing Food Purchases

Partial Systems of Demand Equations with a Commodity Emphasis

THE IMPORTANCE AND USEFULNESS of the complete demand systems
approach for bridging the gap between theory and empirical
analysis of consumer demand behavior are highlighted in the
previous chapters.[1] In this approach, both econometrics and
judgment are used in obtaining elasticity estimates that meet
the restrictions which, in part, are of a theoretical nature.
The complete demand systems approach is primarily concerned with
the structure and estimation of the interdependency of demand
elasticities for various food commodities. However, it is not
necessarily the only way to empirically describe demand behavior.
In the case where interest is in estimating demand for a single
commodity or a small subset of commodities, the alternative of
partial demand systems approaches may be more pragmatic.

The distinction between partial and complete systems
approaches adopted in this chapter arises from the fact that
total budget constraint on utility maximization is generally not
imposed on the partial systems of demand functions. Furthermore,
partial systems demand functions do not satisfy a number of the
theoretical restrictions which were discussed in Chapter 1. The
partial systems approach adopted by many researchers generally
involves ad hoc considerations to cope with more specific
problems since neoclassical demand theory has sometimes not
provided a sufficient a priori basis for empirical demand
analysis. In addition, the partial systems approach may
encompass more than one level of the marketing channel.

More specifically, practical considerations usually limit
empirical applications of a complete system to a few aggregate

---

Contributing authors of this chapter in alphabetic order are
Barry W. Bobst, Chung L. Huang, and Daniel S. Tilley.

commodity groups. This raises questions about aggregation over commodities and the usefulness of the results. In contrast, the partial systems approach often is designed to answer policy questions that are specific to a particular commodity or commodity group.

Considerations concerning the choice of functional form are another basis for justifying the partial systems approach. Empirical application requires the specification of the functional form of the demand equations. Within the context of complete demand systems, the same functional form is generally used for every commodity. There is no general consensus on the issue of the ideal functional form and the particular form used may not be logical for every product. Prais and Houthakker (1955) argue that it is more important to have the flexibility of fitting different forms to different products. If different functional forms are assumed for different commodities, the demand equations cannot be derived from the same utility function.

The ability to incorporate some of the vertical linkages in the farm to a consumer marketing channel is another potential advantage of the partial systems approach. Vertical market structure research has long been popular for agricultural economists. The importance of this area of research stems from being able to analyze the impact of policy variables active at one level in the marketing channel on participants at another level in the marketing channel. This could involve the impact of policy actions at primary producer levels on consumers and/or the impact of consumer policies and programs on primary producer level prices and income. Additional policy variables that act on behavior at intermediary points in the marketing channel could also be encompassed by models of the vertical channel and evaluated in terms of their impact on consumers and producers.

Given the diverse nature of research objectives in dealing with partial systems of demand functions for various food commodities, a comprehensive review and generalization of all studies collected in this chapter appear to be less desirable. This chapter is organized as follows: (1) consumer preference for whole milk versus lowfat milk, (2) dynamic demand analysis of orange juice, (3) structural homogeneity of the beef expenditure pattern, (4) regional demand patterns of broiler meat, (5) a test of disequilibrium hypothesis for beef demand, and (6) conclusions.

CONSUMER PREFERENCE: WHOLE MILK VERSUS LOWFAT MILK

Huang and Raunikar (1979a, 1979b) examined the changes in whole and lowfat milk expenditures based on data collected from the Griffin Consumer Panel. (A description of the Griffin Consumer Panel can be found in Raunikar 1976.) The functional form of Engel curve is a modified hyperbolic function which allows expenditure to increase with income over a range, reach a maximum value, and then decline with further increases in income.

The results of the statistical analyses indicate that race is a significant influence on fresh milk expenditures. In general, white households purchase more fresh milk than nonwhite households, spending about $0.33 and $0.60 more than nonwhite households on whole milk and lowfat milk, respectively.

With regard to household composition, the analyses show that individual members in the household also tend to exert considerable influence on both types of fresh milk expenditures. Particularly, households with younger children appear to purchase consistently more lowfat milk than households with older children. However, a household generally purchases more fresh whole milk than lowfat milk in response to changes in household size.

In terms of the impact of changes in income on household expenditures for each type of milk, it appears that income is of less influence than either race or household composition. The results suggest that household purchases of fresh milk were only slightly responsive to changes in household income. Nevertheless, lowfat milk expenditures were considerably more responsive to income as compared with whole milk at various income levels. Most significantly, the results suggest that household expenditures for fresh milk by type change with the level of income. At low-income levels, whole milk purchases were predominant and were generally preferred to lowfat milk while lowfat milk was generally selected over whole milk and was predominant at high-income levels. Response to income for whole milk expenditures is found to be positive at the low-income level and declines with additional increments of income. In contrast, household expenditures on lowfat milk showed a negative response to increased income at low-income levels and gradually responds positively to income as income further increases. This contrast in expenditure patterns suggests that low-income households may perceive whole milk and lowfat milk differently than high-income households.

On the basis of the present analysis, it can be expected that the market for lowfat milk will continue to grow if the average American family becomes more affluent. Furthermore, it is also apparent that in planning marketing strategies, the dairy industry should take into consideration the effects of various socioeconomic variables on the market demand for whole milk and lowfat milk. The population trends in terms of racial and household composition definitely will exert an influence on the growth of market demand for each type of fresh milk.

DYNAMIC DEMAND ANALYSIS OF ORANGE JUICE

Traditionally, demand theory provides a static interpretation of consumer behavior. The static analysis of consumer orange juice purchasing behavior assumes that the consumer adjusts instantaneously to a new equilibrium when income or prices change. This assumption is somewhat inconsistent with reality in that it ignores adjustments due to home inventories and habit formation over a time horizon. Inventories and habits may result in delays in consumer responses to income or price changes.

Two time-series studies of orange juice demand are reported in this section. The first involves consideration of demand only at the retail level. The second study uses similar demand models in a simultaneous equation model of the orange juice marketing channel.

Retail Purchases

In a study of frozen concentrated orange juice (FCOJ) and chilled orange juice (COJ) consumption, Tilley applied the dynamic flow adjustment model developed by Houthakker and Taylor (1970) to demonstrate the importance of product definition within a commodity group when reaching conclusions about inventory or habit dominance.[2] Empirical analyses were further extended to show the importance of identifying the extent to which the change in consumption is due to changes in number of buyers or purchases per buyer when attempting to regain sales levels. The results are in Table 9.1.

TABLE 9.1. Short-run and long-run responses to prices and length of adjustment period, frozen concentrated orange juice and chilled orange juice

Product-equation (Dependent variable)	Price elasticities			
	Lag coefficient (% change)	Short-run (% change)	Long-run (% change)	Adjustment period[a] (months)
FCOJ				
Per capita consumption (q_t)	1.6748	−1.4385	−0.8589	8
Percentage of families buying (FB_t)	1.8583	−0.8048	−0.4331	7
Purchases per buying household (PH_t)	1.0665	−0.5906	−0.5538	1
COJ				
Per capita consumption (q_t)	0.5074	−0.4300	−0.8474	5
Percent of families buying (FB_t)	0.6397	−0.2893	−0.4522	3
Purchases per buying household (PH_t)	0.7595	−0.2648	−0.3487	3

Source: Tilley (1979, 44), with permission.

[a]The adjustment period is the length of time it takes for 95 percent of the effect to occur. The long-run coefficient is

$$\lim_{J \to \infty} \sum_{j=0}^{J} (1 - \phi)^j \phi \beta = \beta$$

Thus, the adjustment period is the minimum J* such that

$$\frac{\sum_{j=0}^{J*} (1 - \phi)^j}{\sum_{j=0}^{\infty} (1 - \phi)^j}$$

is greater than 0.95 for habit dominance or less than 1.05 for inventory dominance effects.

   The coefficient for lagged FCOJ purchases indicates inventory-type dominance which means that the short-run price elasticity (-1.4) is more elastic than the long-run fully adjusted
price elasticity (-0.86). The length of the adjustment period
is estimated to be eight months.

   For FCOJ, percentage of families buying (FB_t) and purchases
per buying household (PH_t) both exhibit inventory-type responses.
The inventory response is much stronger for FB_t than for PH_t.
This finding identifies the primary source of the inventory-type
response as changes in number of buyers rather than purchases
per buyer. Purchases per buyer increase only slightly when
percentage of families buying increases. It is not possible to
determine whether the consumers' consumption rate also changes
or whether the consumption rate is relatively constant and the
product is actually kept in inventory for consumption in non-
purchase periods.

   Tilley suggests that these results provide an alternative
explanation of inventory-type dominance. The observed phenomenon
may be short-term product switching rather than product stocking. That is, consumers enter and buy then exit, and while they
are not purchasing they may either consume from inventory, cease
to consume, or consume substitute products. On the basis of aggregate per capita purchases, it is not possible to determine
which of the phenomena is occurring. Given the results for FCOJ,
entry and exit rather than product stocking appears to be the
primary source of the inventory-type response.

   For COJ, the coefficients indicate that habit effects domi-
nate with a positive relationship between current and lagged
consumption. Habit effect dominance means that the short-run
price elasticity (-0.43) is less elastic than the long-run
elasticity (-0.85) because the purchase habit takes time to
change. The length of the adjustment period is estimated to be
five months. Percentage of families buying and purchases per
buying household show similar habit-type response properties.

   The two products also differ with respect to the estimates
of the short-run cross-price elasticity. The result suggests
that the COJ prices may have a short-run impact on FCOJ consumption, but the reverse is apparently not true. The long-run
cross-elasticity results are also shown in Table 9.1. For FCOJ
the long-run substitution effects would be lower than the
short-run effects whereas the reverse would be true for COJ.

   The FCOJ and COJ results suggest that the two products'
characteristics are particularly important in the analysis of
short-run and long-run adjustments to price changes. Though the
short-run own-price elasticity for FCOJ was estimated to be 3.3
times greater than the short-run own-price elasticity for COJ,
the long-run FCOJ and COJ elasticity estimates are approximately
equal.

Retail and Wholesale Level Demand for Orange Juice

     In a subsequent study of the FCOJ and COJ market Malick used a similar demand specification in a model of the FCOJ-COJ verti-cal channel.[3] Emphasis in the model is placed on the relation-ships between factors influencing wholesale prices, sales, inventory, retail prices, and consumer purchases. Because raw product producers share in wholesale level revenues through participation plans and cooperative-type agreements, grower returns generally reflect wholesale prices. The complete model is more complex and involves more than consumer demand related analysis which is beyond the scope of this chapter. The consumer purchase (demand) relationships, however, provide the opportun-ity to show how the demand relationships are considered within the context of a model of the vertical structure.

Conceptual model. Conceptually, supplies are treated as exogen-ous to the system because decisions to grow more fruit require more than a five-year planning horizon. Thus, the basic supply-related forces of the model are current inventories and expected future supplies reflected by the current crop forecast. Proces-sors are then hypothesized to set prices such that they attempt to achieve ending inventory objectives. Given processors' pricing decisions, retailers decide how much product to buy and set retail prices. Consumers decide how much of each product to buy. While the explanation appears quite simple, it becomes more complex because feedback between levels is assumed to occur in a short time period which suggests a simultaneous equation system. An additional complicating factor in the FCOJ-COJ relationship is the fact that some frozen concentrated orange juice is shipped in bulk and then reprocessed into chilled orange juice prior to selling to the consumer. Thus, to capture the impact of rapidly expanded bulk movement impact on frozen concentrated orange juice (FCOJ) inventories, the model includes a retail demand relationship for chilled orange juice (COJ), and its impact on bulk FCOJ movement is measured. The seven equa-tions in the model are (1) wholesale price determination, (2) retail-size FCOJ movement, (3) bulk movement, (4) retail price of FCOJ, (5) retail price of COJ, (6) consumer purchases of FCOJ, and (7) consumer purchases of COJ. Two identities were used to define seasonally adjusted wholesale inventory of FCOJ and the price of bulk FCOJ.

Results and implications for demand research. Consumer demand for FCOJ and COJ is influenced by own-price, substitute prices, income, and seasonal component. FCOJ displays a stock effect, while COJ exhibits a habit effect. These results are consistent with the results shown in the single-equation demand analysis in the previous section. However, within the context of the full model, variables which influence retail prices also influence the quantity demanded.

The structural equations suggest that a one-cent change in the retail price of FCOJ results in an opposing change in consumer purchases of FCOJ of 4.1918 gallons per thousand people and a one-cent change in the retail price of COJ implies an opposing 0.5804 gallons per thousand people change in consumer purchases of COJ. Based on the prevailing prices and sales levels, elasticities can be derived. If the mean values for the data period are assumed, own-price elasticities of -1.2212 and -0.3719 for FCOJ and COJ, respectively, are implied. The corresponding long-run elasticities are -1.002 and -0.9904. These results are consistent with those reported in Table 9.1. However, use of the structural equations to calculate elasticities can be very misleading since it does not consider the interdependencies between quantity of FCOJ and COJ purchased and between purchases and retail price.

By considering the reduced-form equations a more accurate understanding of the price-quantity relationship can be achieved. Use of the reduced-form equations allows for the impact of all other endogenous variables on the equation in consideration to be measured. Thus, one cannot truly depict the effect of retail price of FCOJ on quantity of FCOJ purchased via the reduced-form equations because the substitute relationship with chilled orange juice is also measured. This factor should not be viewed as a drawback since the resulting change in FCOJ and COJ consumption will be due not only to the own-price effect but also the substitution effects. Since the reduced-form equations are functions of predetermined values, a roundabout but correct path must be taken to obtain the full effect (own-price and substitution effects) of a change in retail price on quantity demanded. Hence, elasticities derived from the reduced-form coefficients are referred to as total elasticities, whereas those derived from the structural parameters are referred to as partial elasticities.

From the reduced-form equations, the estimated coefficients suggest that a 10 million box increase in crop forecast causes retail price of FCOJ to decrease by 0.383, retail price of COJ to decrease by 1.060, retail purchases of FCOJ to increase by 1.245, and retail purchases of COJ to increase by 0.6465. Given these estimates, the total elasticities that measure a combined price and substitution effects can be derived. Using the mean values of the data period, the above changes correspond to percentage changes of -4.15 for retail price of FCOJ and 4.13 for retail purchases of FCOJ, and -2.60 and 2.52 for retail price and purchases of COJ, respectively. Thus, the total effect of a 1 percent increase in retail price of FCOJ in a month is a 0.995 percent (4.13 divided by 4.15) decrease in retail purchases of FCOJ. Similarly, a 1 percent increase in the retail price of COJ during the month results in a 0.969 percent (2.52 divided by 2.60) decrease in consumer purchase of COJ.

The total elasticities for FCOJ and COJ are somewhat inelastic for the initial month. By considering the lagged effects,

long-run elasticities for FCOJ and COJ are estimated to be
-0.777 and -1.136, respectively. Note that these estimates
suggest a different response pattern for the demand for FCOJ and
the demand for COJ. The results indicate that for FCOJ, the
total elasticity is somewhat less elastic than partial elasticity
derived from the structural equations. For COJ, the opposite is
true. Specifically, the results suggest that in the case of
FCOJ, the substitution effect tends to offset the own-price
effect, while in case of COJ, the substitution effect appears to
enhance the price effect.

A one-cent change in real price of substitutes yields corresponding changes of 2.0689 and 0.1760 in retail purchases of
FCOJ and COJ according to reduced-form impact multipliers. The
corresponding cross-price elasticities based on mean values are
0.9299 and 0.098 during the first month and 0.7459 and 0.284 in
the long run. A $100 change in real disposable income per capita
yields corresponding changes of 1.7436 in quantity of FCOJ and
1.5930 in quantity of COJ according to the reduced-form equations. The corresponding income elasticities, based on mean
values, are 1.3122 and 1.4863 during the initial month and
1.0765 and 3.9582 in the long run.

STRUCTURAL HOMOGENEITY OF BEEF EXPENDITURE PATTERN

Another area of concern in the model specification of income-
expenditure relationship is the question of parametric homogeneity for Engel curves for each socioeconomic class.[3] When economic and socioeconomic characteristics change, policies based
on forecasts of such change cannot be based on parameter estimates from models which implicitly or explicitly assume such
changes cannot occur. Agarwala and Drinkwater (1972) argue that
the familiar Engel curve results require modification when
applied to situations in which the structure of the population
and economy is diverse and changing. Hassan and Johnson (1976b)
examine the parametric homogeneity for Engel curves in Canada
across sample partitions based on cities, family income, life
cycles, age of family head, tenure in home, and education of
family head. Their results, with few exceptions, show that
there is a lack of homogeneity of the Engel curve coefficients
across sample participants. Therefore, meaningful applications
of even the most simple income-expenditure parameters in policy
analysis should be conditioned on evidence of parametric or
structural homogeneity.

The conventional approach to test the assumption of structural homogeneity for Engel curves based on sample partition is not
desirable because it can result in many estimated relationships.
Additionally, partitioning the sample for different socioeconomic
classifications has the limitation of substantially reducing the
degrees of freedom and, hence, reduces the reliability of the
estimates. Using household food purchase data collected from
the Griffin Consumer Panel, Huang and Raunikar (1981) apply an

alternative approach to examine the parametric homogeneity for
Engel curves across household size and household income.
    Spline functions were developed to reflect differences in
income-expenditure relationships by allowing the dependent
variable to take on different functional relationships with
respect to the independent variables in various subintervals of
the domain of the independent variables in a continuous fashion.
Since structural relationships of household food expenditure
were postulated to change among various subintervals of the
independent variables under the spline methods of model specification, this approach can be viewed as an analogy to the varying-
parameter models.
    Four categories of beef expenditures were examined with
separate regression equations. The categories were (1) fresh
beef, (2) ground beef, (3) beef roasts, and (4) steaks. The
results indicate that none of the cubic segments are statistically significant. Furthermore, the results also suggest that
except for ground beef, the quadratic segments for income are
not statistically significantly different from the linear segments. Thus, for ground beef expenditure, the additive quadratic
splines in both income and household size are selected for the
statistically appropriate model. For the other types of beef
expenditures (i.e., fresh beef, beef roasts, and beef steaks),
the statistically appropriate model incorporates a quadratic
spline in household size and a linear spline in household income.
    The estimated relationships for a household size of three
are depicted in Figure 9.1. Note that the patterns of fresh
beef expenditure in response to income differ among income
levels. For example, household expenditures for fresh beef
increase rather rapidly as income increases from $2,000 to
$10,000, remain quite stable between the range of $10,000 and
$24,000, and again increase as household income increases above
$24,000. The standard t-test indicates that the ordinates of
the linear segments are statistically significantly different.
Thus, the null hypothesis of no structural changes can be rejected. This suggests that the marginal propensity to consume is
much higher for the low-income and high-income households as
compared with the middle-income households in the case of fresh
beef.
    The estimated income-expenditure relationships also reveal a
sharp contrast in the expenditure patterns among ground beef,
roasts, and steaks. Household food expenditures for ground beef
reach a maximum approximately at the income level of $8,000 and
then gradually decline as income further increases. The ground
beef expenditure curve tends to rise slightly toward the higher
income levels. This pattern, however, does not seem to be significant. Expenditure on beef roasts, in general, resembles the
ground beef expenditure pattern except for absolute magnitude
differences. Nevertheless, a significant structural change,
unlike that of ground beef, is found at the higher level of
household income. In contrast to ground beef and roasts, the

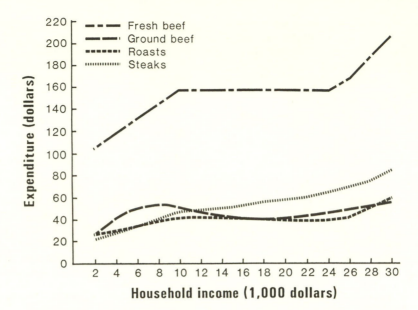

Fig. 9.1. Beef expenditures as a function of household income (household size = 3). (Huang and Raunikar 1981, 109, with permission.)

expenditure curve for steaks shows a steadily increasing pattern as household income increases. Similar to roasts, a significant structural change on the expenditure of beef steaks exists when household income approaches the $25,000 level.

To summarize, the results clearly suggest that beef purchasing behavior changes as household income increases. Households with lower income tend to spend more of their food dollars on ground beef with no appreciable difference between roasts and steaks. As income increases, beef steaks gradually substitute for ground beef. Hence, for the higher-income families, a greater proportion of the budget for beef was spent on steaks with no apparent preference placed on either ground beef or roasts. More specifically, the results suggest that different beef expenditure patterns emerge as household income changes. This implies that at the low-income level, consumers perceive steaks and roasts as one beef item and ground beef as another in allocating their food expenditures on beef. However, as income further increases, their decision process of spending food dollars to purchase beef changes to the extent that they view ground beef and roasts as a beef item in contrast to steaks.

REGIONAL DEMAND PATTERNS OF BROILER MEAT

Demand analyses are often performed at levels of aggregation that are of little use in providing answers to questions about

the effects of exogenous shocks at a lower level of aggregation.[4]
Typically, in a time-series analysis, observations usually aggregate over microunits or households. Thus, the effects of the
variety and complexity of factors influencing individual household behavior tend to average out in aggregate time-series
observations. In a cross-section analysis, it is usually assumed that prices are constant because the data are collected over
a short period of time. Thus, demand relationships in the cross-
section analysis are often specified in terms of socioeconomic
and demographic factors. Since price is often omitted from the
model, direct price and cross-price elasticities cannot be
measured from cross-section data unless utility is additive.
Even in a perfectly competitive market, prices are expected to
vary systematically by quality and region. Furthermore, sales
taxes and transportation costs may result in regional price differences. Using a consumption index and a price index designed
to give a qualitative estimate of the regional variations in
consumption, George and King (1971) observed that all commodities showed some degree of regional variations in their consumption pattern.

Huang and Raunikar (1978) obtain estimates of price and
income elasticities based on pooling time-series and cross-
sectional data of per capita consumption of broiler meat in two
regional markets. Covariance analysis was employed to test the
null hypothesis of no structural change over time and of overall
homogeneity between the two regional markets. The results suggest that one regression line would not appropriately represent
the demand relations for the entire set of observations. Thus,
models were developed to allow for differences in broiler meat
demand patterns due to geographic location.

Estimates of price and income elasticities for broiler meat
for 22 cities in north central-western and eastern-southern
markets for 1972-75 are presented in Table 9.2. In general, the
estimates appear to be consistent with a priori expectations and
available elasticity estimates of previous studies (Brandow
1961; George and King 1971; Hassan and Johnson 1976a). However,
caution should be exercised in any attempt to compare and
interpret various measures of elasticity. Meaningful statements
about the estimated relative responses can only be made within
the context of the model and the data employed.

TABLE 9.2. Estimates of price and income elasticities for broiler
meat, pork, and ground beef at retail level

Variable	Elasticities	
	North central-western market	Eastern-southern market
Price		
Broiler meat	-.170	-.169
Pork	.320	--
Ground beef	.132	.070
Income	.150	.144

Source: Huang and Raunikar (1978, 591), with permission.

The estimates of demand elasticities from this study seem to be lower than most of the previous studies. The results may be attributed in part to the fact that the average price paid for broiler meat is appreciably below that of other meat products due to the greater technical efficiencies gained in the production sector of the poultry industry.

Although there were considerable similarities in the estimated elasticities, it appears that the consumption patterns of broiler meat between the two markets differ considerably. Covariance analysis suggested that the demand relations for the two markets are not homogeneous. Specifically, factors that affect the level of broiler meat consumption in these two regional markets may be different. The hypothesized model accounted for the variation in the consumption of broiler meat in the eastern-southern market quite well but not for the north central-western market. Pork was considered a strong substitute for broiler meat in the north central-western market while its effect upon broiler meat consumption in the eastern-southern market could not be ascertained. Ground beef, as expected, competed strongly with broiler meat in both markets. In addition, consumption responses with respect to racial distribution are also quite different in the two markets. It is estimated that cities in the eastern-southern market having a 1 percent higher than average white population, other things being constant, demand an average of .83 percent less broiler meat per capita than those cities with mean distribution of white population. The corresponding estimate for the north central-western market, nonetheless, was only .16 percent.

TEST OF THE DISEQUILIBRIUM HYPOTHESIS FOR BEEF DEMAND
        The theory of market equilibrium holds that prices are sufficiently flexible to re-establish equilibrium within the time period of observation in response to shifts in demand and supply functions.[5] However, many factors could prevent prices from being sufficiently flexible to re-establish equilibrium in each period as required by the theory. Prices could be administered, either by regulatory agencies or by oligopolistic price leadership or formula arrangements and subject to change only after considerable delays (Phlips 1980). In such a case, and given frequent shifts in demand and supply functions, markets could exist in a more or less permanent state of disequilibrium.
        Markets in disequilibrium have important implications for the interpretation of time-series quantity data. As suggested by Fair and Jaffee (1972), observed quantity will be the minimum of either quantity demanded or quantity supplied in a strictly flow market.
        Interpreting quantity data in markets which are in disequilibrium presents some seeming paradoxes, particularly when quantity demanded (disappearance) is estimated as a residual. Beef is an example of a commodity which has these characteristics.

Monthly U.S. beef disappearance is approximately equal to domestic beef output plus imports with very minor adjustments for exports, stocks change, and nonmarket consumption. Supposing for the moment that beef, or a commodity with stocks-flow characteristics like beef, is usually in a state of disequilibrium, interpretations of production and disappearance data would be affected.

First, suppose that the market is in an excess demand state. Price would be adjusting upwards, but the adjustment would be insufficient to establish equilibrium during the observation period (a month). In this case the observed transactions level would be governed by supply. The measured output level would be an unbiased estimate of quantity supplied.

Demand governs the observed level of transactions if the market is in an excess supply state. In this case, and again assuming no measurement error, the residual estimate of disappearance would be an unbiased estimate of quantity demanded at the prevailing price. However, the direct observation of production would be less than the quantity producers would otherwise be willing to supply at the prevailing price.

The data biases previously discussed will cause inconsistent estimation of demand and supply functions' parameters if conventional econometric methods are used. A test of the hypothesis of disequilibrium effects in monthly beef data for the period 1966-79 has been conducted. The model included a demand equation, a supply equation, a price change function, and a quantity measure which specified that quantity traded is the minimum of supply and/or demand. A two-stage least-squares procedure, modified to include the cross-equation constraint, was used to estimate the model. The results imply that the rate of change of retail beef prices is too slow to establish equilibrium within the one-month time period of observation. Thus, the analysis suggests that disequilibrium is a factor that needs to be taken into account in evaluating monthly beef data.

Equilibrium versions of these demand and supply functions were estimated for purposes of comparison. These functions omit the price change variable since in the equilibrium model all price effects are presumed to be represented in the price variable itself. Parameter comparisons disclose that the disequilibrium model's parameters have the same signs as their equilibrium model counterparts and are quite similar in value for many variables. The largest differences in parameter estimates are for retail price and for the steer/cow price ratio in the supply function.

These results suggest that quantity data from markets in disequilibrium require careful interpretation. Under disequilibrium conditions, disappearance data can at times be interpreted as measures of quantity demanded, but at other times they can not. Although market disequilibrium appears to offer unorthodox interpretations of market data, the concept has been shown to have some plausibility, at least in the case of beef. Given the increasing emphasis on short-run econometric forecasting of commodity markets, attention needs to be given to the problems raised by the disequilibrium hypothesis for the interpretation and analysis of time-series data.

CONCLUSIONS
This chapter has reviewed some contributions in the area of partial systems of demand equations. The models developed in this chapter represent a variety of single- and multiple-equation demand analyses concerning the functional form and structural homogeneity of the Engel curves and the dynamic nature of consumer behavior.

While research efforts reflected in this chapter are quite extensive, they are primarily problem oriented with respect to correct model specification relative to the research objectives, data availability, and estimation methods. Thus, an important attribute of this chapter has been the shift from estimating parameters of demand structures to the estimation of certain particular demand parameters for purposes of prediction and policy analysis. Estimates of some key parameters, from a pragmatic point of view, are important objectives of applied economics.

NOTES

1. Empirical results and discussion presented in this section are adapted from Tilley (1979).
2. Empirical results and discussion presented in this section are adapted from Malick (1980) and contributed by Tilley.
3. Empirical results and discussion presented in this section are adapted from Huang and Raunikar (1981).
4. Empirical results and discussion presented in this section are adapted from Huang and Raunikar (1978).
5. Contributed by Barry W. Bobst.

REFERENCES

Agarwala, R., and J. Drinkwater. 1972. Consumption functions with shifting parameters due to socioeconomic factors. Rev. Econ. and Stat. 54:89-96.
Barten, A. P. 1977. The systems of consumer demand functions approach: A review. Econometrica 45:23-51.
Brandow, G. E. 1961. Interrelations among demands for farm products and implications for control of market supply. Pa. Agric. Exp. Stn. Bull. No. 680.
Brown, J. A. C., and A. S. Deaton. 1972. Surveys in applied economics: Models of consumer behavior. Econ. J. 82: 1145-236.

Fair, R. C., and D. M. Jaffee. 1972. Methods of estimation for
     markets in disequilibrium. Econometrica 4:497-514.
George, P. S., and G. A. King. 1971. Consumer demand for food
     commodities in the U.S. with projections for 1980. Giannini
     Found. Monogr. No. 26, Univ. of California, Berkeley.
Hassan, Z. A., and S. R. Johnson. 1976a. Consumer demand for
     major foods in Canada. Agric. Can. (Ottawa), Econ. Br.
     Publ. 76/2.
_____. 1976b Family expenditure patterns in Canada: A sta-
     tistical analysis of structural homogeneity. Agric. Can.
     (Ottawa), Econ. Br. Publ. 76/3.
Houthakker, H. S., and L. D. Taylor. 1970. Consumer demand in
     the United States 1929-1970. 2d ed. Cambridge, MA: Harvard
     Univ. Press.
Huang, C. L., and R. Raunikar. 1978. The demand for broiler
     meat. Poult. Sci. 57:588-92.
_____. 1979a. Consumer preference: Changes in fresh milk con-
     sumption. Am. Dairy Rev. 41:66A-66B.
_____. 1979b. Household expenditures of whole and lowfat milk:
     Implications for milk marketing. J. Food Distrib. Res.
     10:25-30.
_____. 1981. Spline functions: An alternative to estimating
     of income-expenditure relationships for beef. South. J.
     Agric. Econ. 13:105-10.
Malick, W. M. 1980. A simultaneous equation model of the
     Florida retail orange juice marketing system. Thesis, Univ.
     of Florida.
Nerlove, M. 1958. Distributed lags and demand analysis for
     agricultural and other commodities. USDA, AMS, Handb. No.
     141. Washington, DC.: U.S. Government Printing Office.
Phlips, L. 1980. Intertemporal price discrimination and sticky
     prices. Q. J. Econ. 94:525-42.
Prais, S. J., and H. S. Houthakker. 1955. The analysis of
     family budgets. Cambridge: Cambridge Univ. Press.
Purcell, J. C., and R. Raunikar. 1967. Quantity-income elastic-
     ities for foods by level of income. J. Farm Econ.
     49:1410-14.
Raunikar, R. 1976. The Griffin consumer research panel: Estab-
     lishment and characteristics. Ga. Agric. Exp. Stn. Res.
     Rep. No. 232.
Thomas, W. J. 1972. The demand for food. Manchester: Man-
     chester Univ. Press.
Tilley, D. S. 1979. Importance of understanding consumption
     dynamics in market recovery periods. South. J. Agric. Econ.
     11:41-46.
Waugh, F. W. 1964. Demand and price analysis: Some examples
     from agriculture. USDA, ERS, Tech. Bull. No. 1316.
     Washington, DC: U.S. Government Printing Office.

CHAPTER 10

Socioeconomic, Demographic, and Psychological Variables in Demand Analyses

ECONOMISTS have long been interested in incorporating socio-economic and psychological variables in demand analyses. Much of the incentive came from policymakers charged with making welfare programs more effective. Welfare comparisons provide the rationale for different treatments of family types on income tax schedules, income maintenance programs, or government food programs. Researchers can now apply an expanded set of methods to comprehensive data sets in their search for previously undis-covered relationships. Potential benefits include increased understanding and knowledge of how demand for specific commodities may be influenced by socioeconomic and psychological characteristics.

The impact of sociodemographic and psychological variables on food consumption has been analyzed mainly from cross-section-al data. Cross-sectional data usually exhibit minimal price variations. They reduce the use of traditional economic theory since "adding-up" becomes the only restriction left that can be used in empirical work. Typically, a traditional demand func-tion is then specified as a function of income. However, pre-vious research has shown that income alone has a low explanatory power in the analysis of cross-sectional data, especially with disaggregated consumption (Hassan and Johnson 1977). This find-ing suggests that sociodemographic variables, which usually exhibit large variations in cross-sectional data, play a key role in explaining consumption behavior.

The introduction of sociodemographic variables in demand analysis has presented new challenges for both theoretical and applied research. One question concerns the usefulness of con-sumer theory when such variables are considered. The economic unit of consumption is typically the household, with a given household income distributed to each household member. Very little work has been done on household, as distinct from individ-ual, preference functions. It is usually assumed that the household is a closely knit group that behaves in such a way that justifies the existence of a household preference function (Samuelson 1966). It is usually argued that households' preferences are functions of their sociodemographic characteristics, reflecting the fact that households with different characteristics have different physical and psychological needs. Thus, in the absence of price variations, demand functions can be specified as

$$x_i = x_i(M, N, S) \qquad (i = 1, \ldots, n) \qquad (10.1)$$

where x_i is the family consumption of the ith commodity, M is household income, N represents family size, and S is a set of other sociodemographic or psychological variables such as family assets, education, occupation, location, age–sex family composition, marital status, and so forth.

Although it is clear that theoreticians have made considerable progress in incorporating sociodemographic variables into consumer theory during the last decade (Barten 1964; Brown and Deaton 1972; Muellbauer 1974, 1980; Kakwani 1977, 1978; Lau et al. 1978; Pollak and Wales 1978, 1980), the importance of these variables has been emphasized by applied researchers for more than fifty years (Engel 1895; Sydenstricker and King 1921; Prais and Houthakker 1955).

Applied research concerning the effects of sociodemographic variables on consumption behavior can be divided into two broad categories. In the first category, the objective of the research is to explain and predict consumption with no attempt to measure welfare. In this case, the research usually consists of estimating the parameters of directly specified demand functions. In the second category, the objective of the research is to compare the well-being of households with different sociodemographic profiles. This has been done by estimating equivalent scales or adult equivalent scales, which can be interpreted as cost-of-living indices.

The estimation of the equivalent scales for welfare analysis has received much attention over the years. Defining these scales based on nutritional requirements has appeared attractive for most food commodities (Cramer 1971). For example, the U.S. poverty standards are determined from a food scale derived from economy food plans of different household types based on considerations of nutrition and palatability (Orshansky 1965; Seneca and Taussig 1971). However, such nutritional scales are based on normative judgments rather than on consumer behavior. Another approach is that equivalent scales should be estimated based on consumer behavior (Prais and Houthakker 1955). The Prais and Houthakker model has been widely used in applied research (Nicholson 1949; Forsyth 1960; Price 1970; Singh and Nagar 1973; McClements 1977; Muellbauer 1977, 1980). However, Muellbauer (1975, 1980) has pointed out some limitations of the Prais and Houthakker model. He argues that the equivalent scales are not identified and that they give a proper measure of cost of living only when the model is derived from a Leontief utility function, implying a zero substitution restriction. Other examples of estimation of equivalent scales can be found in Kapteyn and Van Praag (1977) and and Bojer (1977).

Pollak and Wales (1979) argue that the validity of equivalent scales for welfare analysis depends upon whether or not the sociodemographic variables are subject to control by the household. While some variables (race, sex, etc.) are not controllable, some others (number of children, location, etc.) can be chosen by the household. Pollak and Wales assert that welfare analysis should be conducted based on unconditional preferences, that is, on preferences that do not depend on sociodemographic variables subject to control.

Although much work has been done on the welfare implications of selected sociodemographic variables, most of the research on the effects of these variables on demand was conducted to explain and predict consumption behavior. In this case, the researchers were most interested in estimating parameters that measure the total effect of sociodemographic variables rather than identifying equivalent scales. This approach has been very common in applied research (Buse and Salathe 1979; Price et al. 1980).

The following sections of this chapter describe the results of four research efforts that have studied relationships between demographic and psychographic variables and food purchase decisions. A final section develops generalizations that are supported by these research efforts. Areas where additional research efforts are needed are also identified.

SOCIOECONOMIC CHARACTERISTICS AND HOUSEHOLD EXPENDITURES

Models that attempt to isolate the net impact of income and other socioeconomic characteristics on household expenditures are reported in this section.[1] The ultimate objective of the research is to explain and predict in an accurate way household expenditure patterns for the United States using models that evaluate the influence of existing and proposed legislation on household income and on food consumption. The results will also be of considerable interest to economists attempting to predict future food price and consumption movements based on changes in the socioeconomic characteristics of the domestic population.

Engel Functions

The results are primarily from data collected in the 1960-61 Bureau of Labor Statistics Consumer Expenditure Survey (1960-61 BLS CES) and the 1955 and 1965 U.S. Department of Agriculture Household Food Consumption Surveys (USDA HFCS). Although part of the differences observed in the results may reflect the fact that price levels have changed dramatically over this period causing changes in consumer demand, the differences discussed here are for households varying by race, location, size and composition, income, and other characteristics for which data were available in the surveys.

Equilibrium status. Motivated by the results of Crockett and Friend (1960), Buse and Salathe indicated in their preliminary analysis that many households had recently experienced changed circumstances: people were widowed, married, and divorced, changed residences from one city to another or from rental units to purchased units; children left home for the armed services, to attend college, and so forth. The researchers hypothesized that these changes in status take time to be fully reflected in household expenditure patterns. Thus, when such changes had taken place in the past year, the households were classified as being in "disequilibrium," that is, not having had sufficient time to fully adjust their expenditure patterns. Their analysis also indicated that "equilibrium" and "disequilibrium" households responded differently depending upon whether they were renting or owning their living accommodations. They subdivided the BLS CES households into four subgroups: (1) homeowners in equilibrium, (2) homeowners in disequilibrium, (3) renters in equilibrium, and (4) renters in disequilibrium.[2] Engel curves for fourteen expenditures categories were estimated for each of the four groups of households.

Adult equivalent scales. In both the 1955 and 1965 Household Food Consumption Surveys, Buse and Salathe (1979) found that sex and age of household members were statistically significant in explaining household food expenditure behavior. In the 1955 data, sex and age of household members (the adult equivalent scale) were important in explaining all household food expenditures except fruits. The coefficients for number of adult equivalents (A) were at least twice their standard error for every food group. Furthermore, the coefficients of the number of adult equivalents squared (A^2) were negative and at least twice their standard error for every food group except vegetables and grain products.

The parameter estimates on income and the adult equivalent scales for the 1965 USDA HFCS are summarized in Table 10.1 where average household coefficients for the Engel function are given. The negative sign of the coefficient of the number of adult equivalents squared (A^2) indicates that household food expenditures increase at a decreasing rate as the number of adult equivalents in the household increases.

Overall, beef and pork, and dairy products exhibit the least response to changing household size. The greatest response to an increase in number of adult equivalents is in fruits. Buse and Salathe (1979) also tested nine hypotheses on the assumptions incorporated into the continuous adult equivalent scales. Gener-

TABLE 10.1. Average household coefficients for Engel function, 1965 USDA HFCS

Expenditure	Constant[b]	Coefficients[a]				
		Y	$Y^2 \times 10^{-3}$	A	A^2	$AY \times 10^{-2}$
Total food	5.573	0.0490	−0.0500	7.598	−0.281	0.504
Vegetables	0.741	0.0045	−0.0035	0.902	−0.031	0.022
Grain products	0.706	0.0030	−0.0035	0.914	−0.024	0.025
Beef and pork	1.271	0.0153	−0.0141	1.623	−0.048	0.139
Dairy products	0.653	0.0041	−0.0047	1.090	−0.052	0.038
Fruits	0.200	0.0060	−0.0065	0.489	−0.019	0.033

[a]Coefficients can be interpreted as dollar per week change in household expenditures associated with a one-unit change in the corresponding independent variable adult equivalent scale (A) or income (Y).

[b]The constant term assumes that households consumed all their meals at home (M = 1).

ally, the scales conformed to prior expectations. For example, adult females consumed less total food, beef and pork, grain products, and milk, and consumed about the same amount of vegetables and fruits as adult males. With the exception of fruits, elderly persons spent less on food than adults aged 20-55. Children were found to consume less total food, vegetables, beef and pork, and more grain products than adults.

In the 1965 BLS CES data, Buse and Salathe (1979) found that household size and composition were more important in explaining household expenditures on food consumed at home than on any other household expenditure. Furthermore, for some expenditures, household size and composition showed substantial variation across types of households (homeowners and renters).

Results generally conform to a priori expectations. The results suggest that the addition of an adult male to the household will cause some household expenditures to increase and others to decline in order to balance the household budget. An adult male has a positive impact on food consumed at home, alcohol and tobacco, transportation, recreation, and education and reading expenditures; a negative impact on household operations; and mixed impact on the remaining expenditures.

Under equivalent assumptions, the addition of an adult female to the household causes expenditures on food consumed at home, household utilities, clothing, medical care, personal care, and education and reading to increase while household expenditures on food consumed away from home, alcohol plus tobacco, and household operations decline.

Children increase household expenditures for food consumed at home, clothing, transportation, personal care, and education while the addition of an adult over 55 increases household operation expenses and medical care costs at the expense of the remainder of the household budget.

Buse and Salathe calculated the theoretical budgets for households of various sizes and different compositions within any given size.[3] The results indicate that the level of expenditures on food consumed at home increases with household size and age of children in each of the household tenure types. The household's expenditures on food consumed at home decreases for homeowners, but increases for renters as the age of adults increases. In contrast, expenditures on food consumed away from home declines with the age of adults and usually does not increase with size. An increase in the average age of children in the household usually does not cause a corresponding increase in expenditures on food consumed away from home.

Food expenditures per adult equivalent also depend upon the location of the household (region and urbanization). Ceteris paribus food expenditure per adult equivalent are highest for households located in the Northeast and lowest for households located in the South. In addition, rural nonfarm households spend less per adult equivalent on food than either rural farm or urban households. Urban residents have the highest average expenditures on total food, grain products, and beef and pork, while rural farm residents spend more on vegetables, dairy products, and fruits. Food expenditures also are lower for an average black household than for an average white household.

In general, they found the level of expenditures to be closely related to the age of adults and children and to the size of the household. Thus they conclude that household size was an important factor explaining household expenditures but that a simple per capita specification was not a satisfactory method of incorporating size and composition into the Engel functions.

Income and Sociodemographic Characteristics

In addition to income and household composition, Buse and Salathe examined a large list of sociodemographic characteristics in both the 1960-61 BLS CES and the 1965 USDA HFCS. They found that in their overall tests of significance, the sociodemographic characteristics are important in explaining household expenditures. Their coefficient of determination ranged from .30 to .60 for the 5,592 observations in the 1965 USDA HFCS and from .10 for miscellaneous expenditures to .60 for clothing and food consumed at home for homeowners in disequilibrium in the 1960-61 BLS CES.

There is also clear evidence of different parameters for households of different educational levels, race, occupation, employment status, and region in both data sets. Because of the interaction variables included in the analysis, the number of variables included, and the wide range of expenditures for which they estimate Engel functions, it was not possible to summarize the effect of each variable separately. Buse and Salathe described the results in terms of the impact of the socioeconomic variables on the marginal propensities to spend and the expenditure elasticities.[4]

Expenditure elasticities. The calculated average sample and partial income elasticities for the 1965 USDA HFCS are tabulated in Table 10.2.[5]

TABLE 10.2.   Sample and partial income elasticities related to household head's
education, employment, race, income, and household size, 1965 USDA HFCS

Characteristic	Total food	Vegetables	Grain products	Beef and pork	Dairy products	Fruits
Constant	.2005	.1512	.1047	.2827	.1322	.2990
Income	−.000409	−.000235	−.000244	−.000521	−.000303	−.000648
Adult equivalents	.021	.007	.009	.026	.017	.016
Education						
< 8 years	.086	.057	−.011	.035	.039	.015
8–11 years	.015	.027	.007	.035	.039	−.040
12–15 years	−.025	−.027	−.004	−.011	.007	.010
16 or more years	−.022	−.003	.010	.039	−.020	.005
Employment						
Employed	−.028	−.030	−.024	−.026	−.032	−.065
Not employed	.012	.013	.010	.011	.013	.026
Race						
White	−.014	−.020	−.003	−.022	−.010	−.025
Black	.118	.165	.014	.148	.061	.209
Other	−.064	−.084	.063	.085	.055	.134

Note: Based upon equations in table evaluated at average income and expenditures
for all households.

The estimated expenditure elasticities over the entire sample
of households are 0.226, 0.154, 0.110, 0.297, 0.146, and 0.301
for total food, vegetables, grain products, beef and pork, dairy
products, and fruits consumed at home, respectively (see Buse
and Salathe 1979).   Comparing the estimated elasticities with
those obtained by George and King (1971) reveals that they are
moderately lower for total food, vegetables, and dairy products,
higher for grain products and fruits, and about the same for
beef and pork.   The differences are probably attributable to the
more detailed specification of the Buse and Salathe Engel functions.   There is ample evidence that expenditure response to a
change in income is conditioned by race, education, household
size and composition, and location.

The Buse and Salathe coefficients are additive.   This makes
it easy to calculate the elasticity for any particular household
once its characteristics have been specified.   They are calculated by multiplying the average marginal propensities in Table
10.2 by the appropriate reciprocal of the proportion of income
spent on each food group.   They can be interpreted as the average
elasticity for all sample households and the partial adjustments
required to adjust that average to a particular set of socioeconomic characteristics (Table 10.2).   For example, the expenditure elasticity of total food for a white family of 3.55 adult
equivalents whose female head had less than eight years of
education, was not employed outside the home, and had an average
weekly income of $120 would be calculated as follows:

0.2005 − 000409(120) + 0.021(3.55) + 0.086
− 0.028 − 0.014 = 0.270

Thus, the coefficients (Table 10.2) can be interpreted as the
impact of a particular household characteristic on the average
elasticity for all households.

The income elasticities exhibit considerable fluctuation

depending upon the households' characteristics. For example, the income elasticity for food for a black household consisting of four adult equivalents whose head has less than eight years of education and is not employed outside the home is 0.45. This compares to 0.20 for a white household consisting of two adult equivalents whose household head has eight to eleven years of education and is employed outside the home. Thus, household sociodemographic characteristics are essential to explaining food purchasing behavior.

Buse and Salathe also calculate the impact of the sociodemographic characteristics of the household in the 1960–61 BLS CES as adjustments to the overall income elasticities. In Table 10.3 estimated average expenditure elasticities for all families in each household group are given.

TABLE 10.3. Estimated average expenditure elasticities for all families in each household group, 1960–61 BLS Consumer Expenditure Diary Survey

| | Household classification | | | | |
| | Equilibrium households | | Disequilibrium households | | All |
Expenditure category	owners	renters	owners	renters	households[a]
Food consumed at home	0.268	0.419	0.278	0.362	0.320
Food consumed away from home	0.942	1.086	1.051	0.817	0.998
Alcohol plus tobacco	0.595	0.828	0.695	0.767	0.736
Shelter plus other real estate	0.563	0.532	0.593	0.551	0.584
Household utilities	0.292	0.328	0.315	0.352	0.308
Household operations	0.863	0.860	0.885	0.974	0.903
Household furnishings and equipment	0.768	0.724	0.892	0.695	0.782
Clothing	0.799	0.926	0.752	0.868	0.830
Transportation	0.694	0.856	0.766	0.766	0.753
Medical	0.460	0.612	0.567	0.731	0.574
Personal care	0.498	0.695	0.604	0.686	0.614
Recreation	0.868	0.882	0.761	0.832	0.838
Education plus reading	0.590	0.770	0.450	0.620	0.588
Miscellaneous	1.011	1.104	0.642	1.165	0.943
Total expenditures	0.591	0.694	0.608	0.663	0.637

[a]Weighted average (grand mean) for all households used in the analysis.

The elasticities for the 14 expenditure categories represent the average elasticities for all households of a given tenure type. The estimated expenditure elasticity for food consumed at home ranged from 0.268 for homeowners in equilibrium to 0.419 for renters in equilibrium. The elasticity for food consumed away from home ranged from 0.817 to 1.086 across the four tenure types. The lowest elasticity was for disequilibrium renters and the highest for equilibrium renters. Generally, households who were homeowners had the lowest expenditure elasticities and renters the highest.

In each of the four groups, the expenditure elasticity for food consumed at home increased with the level of education of the household head. The elasticity for food consumed away from home over each of the four subsamples ranged from 0.817 to 1.086. In addition, these elasticities declined as the education of the

household head increased, were higher for urban households than
rural households, were higher for households residing in the
Northeast than for households in the South, and were lower for
black households than for white households.

INCORPORATING PSYCHOLOGICAL AND SOCIAL FACTORS
    This section emphasizes the effects of psychological and
social factors as well as demographic characteristics on consumption patterns.[6] Specific consumer expenditures, as with all
behavior, are motivated by both needs and values. The need
level at which an individual operates will affect consumption.
Self-actualization theory, as first expounded by Goldstein
(1939) and redefined by Maslow (1970), views all behavior as
motivated by a progressive level of needs. These needs move
from basic, deficiency-oriented needs up through higher, growth-
oriented needs. It is hypothesized that consumption of food
represents the most basic need, a physiological need, and that a
higher value for this need is directly related to consumption of
those foods which are basic to survival. However, since food
can represent a multiplicity of factors to individuals, the
specific food items consumed can be related to any need level.
Cultural factors, ethnic background, geographic origin, and
early family life style and relationships can combine to associate certain food or food patterns with certain need levels.
    The following results stem from two data sources: (1) the
1973-74 U.S. Bureau of Labor Statistics' Consumer Expenditure
Survey (BLS CES) and (2) a 1972-73 Washington State survey of 8-
to 12-year-olds and their families.[7] The results of the analysis
of nine sets of variables using one or both of these sets of
data are subsequently given. Under each set of variables there
are two categories: (1) the effects of total household food
expenditures and (2) the effect on specific food items.
    The results for total household food expenditures are from
the BLS CES data and from the Washington State data (West and
Price 1976; and West et al. 1978). The results from the BLS CES
data include observations on 9,794 households that did not purchase food stamps. Food stamp recipients were excluded because
it is hypothesized that they do not react to the explanatory
variables to the same degree as nonparticipants because of the
restrictive provisions of the program.

Results for Specific Socioeconomic Variables
Race. The Washington State data showed no significant effect of
race on total food expenditures for either blacks or chicanos.
Likewise, the 1973-74 BLS CES data showed no significant effect
between blacks and nonblacks.
    Within food categories, results from the 1973-74 BLS CES
data show significantly lower expenditures by blacks for cereal
and bakery products and dairy products, but significantly higher

expenditures for meat, poultry, and fish. No significant differences were found for expenditures on fruit or on vegetables.

Analyses of the Washington State data showed black children
to consume significantly lower amounts of fluid milk, major
dairy products, fresh vegetables, and bread and rolls than
whites but significantly larger quantities of all meat, cooked
cereal, and nonalcoholic beverages than whites. Chicano children
consumed less beef, fresh vegetables, total vegetables, ready-
to-eat cereal, sweets, jams and jellies, and fats and oil than
white children but more cooked cereal and nonalcoholic beverages.

The Washington State household data showed significant
racial differences in the frequency of serving many major food
groups. The frequency of serving dried vegetables, eggs, juice,
and rice was higher among black households than white while the
frequency of serving milk (to adults) and potatoes was lower.
Chicano households served dried vegetables, juice, eggs, and
rice more frequently than white households but they served
nondried vegetables, meat or fish, milk (to adults), and potatoes
less frequently. There were no significant racial differences
with respect to the frequency of serving fruit. Therefore, even
though race has little effect on total food expenditures, significant differences exist with many food items.

Household size. The Washington State data show significant
economies of size even after using unit equivalent scales which
incorporate economies of size found in the 1955 USDA household
budget data (Price 1970). There were differences by ethnic
group with Chicano households showing more economies of size
than black households which in turn showed more economies of
size than white households.

Using the same set of unit equivalent scales, the 1973-74
BLS CES data did not consistently show significant economies of
size for the five major food groups analyzed. Vegetables were
the exception. The results for the fruit and meat categories
were inconclusive. These findings indicate that economies of
size are much greater among some groups than others. Both
purchasing practices and the types of food served may vary with
household size.

Consumption data for the 8- to 12-year-old Washington State
child showed household size to be significantly negative for
fluid milk, fresh fruit, and all fruit. This result gives some
evidence that food consumption patterns vary with household size.

In a study of the types and variety of fruits and vegetables
served by the household with the Washington State data, household size was positively related to the variety of items served.
This finding supports the hypothesis that with a greater number
of persons who have a greater variety of preferences, a greater
variety of fruit and vegetable items are served.

Geographic origin and region. The 1973-74 BLS CES data show
food expenditures to be roughly 150 percent higher in the North-

east than in other regions of the country. Some of the difference may be due to higher prices and/or the consumption of more
expensive foods.

The results from the BLS CES data showed significant
regional differences for all five categories of food expenditures. For cereal and bakery products, and for meat, poultry,
and fish, expenditures in the Northeast were significantly
higher than in the other regions. Expenditures for dairy products were highest in the Northeast with the West also being
significantly higher than the other two regions. Fruit expenditures in the West and in the Northeast were roughly equal and
significantly higher than in the other two regions. Expenditures for vegetables were highest in the Northeast and lowest in
the northcentral region.

The Washington State data included information on the region
where the parents were raised. This variable showed that the
child's consumption of fluid milk and dairy products were lower
for those of southeastern origin than for the children whose
parents came from other regions. Consumption of beef, all meat,
fresh vegetables, and all vegetables was higher for children
whose parents were raised in the northeast/central regions than
from other regions (due to the low numbers of observations,
northeast and central was a single region). These children also
consumed less cooked cereal.

Rural–urban and city size. The 1973–74 BLS CES data show households in the large SMSAs (population of one million or more) to
spend significantly more for total food than households in other
areas. Expenditures were roughly 7 percent higher in the large
SMSAs. Again, differences can be due to higher prices and/or
the use of more expensive foods.

The BLS CES expenditure data showed significant differences
by size of SMSA for four of the five categories of food. Cereal
and bakery products showed no significant differences. Expenditures for meat, poultry, and fish were 11 percent higher in the
large SMSAs than in the small- and medium-sized SMSAs and 16
percent higher than in the non-SMSAs. Fruit expenditures were
roughly 11 percent higher in the large SMSAs than in other
areas. Expenditures for vegetables were 11 percent higher in
the large SMSAs than in the small SMSAs and the non-SMSAs but
only about 5 percent higher than in the medium-sized SMSAs.
Dairy product expenditures showed a different pattern perhaps
because of the type of institutionalized pricing for fluid milk
under the various state and federal marketing orders. The
higher expenditures occurred in the largest SMSAs and in the
non-SMSAs. Expenditures in the medium and small SMSAs were 8
percent and 5 percent, respectively, under those in the large
SMSAs.

The Washington State data on children showed no clear patterns of differences in consumption by size of city. Some differences by geographic area within the state did exist, suggest-

ing variations in consumption patterns within relatively small
geographic areas. Since the state is relatively diverse with
respect to climate, topography, type of agriculture, concentration of population, and population growth, more diversity within
the state exists than with most geographic areas of the same
size.

Life cycle. The analysis of the life cycle categories with the
BLS CES data showed significant differences in expenditures among
the various stages even though expenditures were on an adult
equivalent basis and the log of household size was included in
the model. The life cycle stages were: Stage 1-- one- and two-
member households with the head less than 35 years of age or
three-or-more-person households with the head less than 30 years
of age; Stage 2--three-or-more-person households with the head
over 30 years of age; Stage 3--one-or-two-person households with
the head 35-64 years of age; Stage 4--one-or-two-person households with the head 65 years and over.

The middle-aged Stage 3 households had the highest food
expenditures on an adult equivalent basis. The second highest
expenditures occurred with Stage 4 households with heads over 65
years of age. Next in this hierarchy was the Stage 2 household.
The stage with the lowest food expenditures was the one with the
head generally less than 35 years of age. For the typical family, adult equivalent food expenditures would be lowest before
children, increase with the advent of children, increase again
when the children leave home, and then modestly decline after
retirement. Since the analysis is based on cross-sectional
data, this assumes that generational differences are small or
nonexistent.

The same four life cycle stages will be used for specific
types of foods. The young Stage 1 household had significantly
lower expenditures of all five food categories than households
in Stage 2. The middle-aged Stage 3 households had higher expenditures for cereal and bakery products and for dairy products
than did households in Stage 2. The Stage 4 households with
elderly heads had higher expenditures for cereal and bakery
products, dairy products, and vegetables than households in
Stage 2.

Over the typical family life cycle, the lowest expenditures
per adult equivalent for the young households without children
are followed by an increase in all five categories of food with
the advent of children (Stage 2). After the children leave home
(Stage 3), increases in per-adult equivalent expenditures occur
for cereal and bakery products and for dairy products. At Stage
4, decreases in expenditures on cereal and bakery products and
on dairy products occur along with a slight increase in expenditures for vegetables.

Income. The Washington State data showed a relatively low-income
elasticity of 0.06 for the total sample and nonfood stamp recip-

ients. Results from the 1973-74 BLS CES data indicate a higher elasticity of 0.13 for all nonfood stamp recipients. The Washington State data give some evidence of income elasticity differences among different groups. For the total sample, it was highest among chicanos (0.08) and lowest among whites (roughly zero). Blacks were intermediate at 0.06. Since the chicano sample had the lowest income levels, the higher elasticity may be attributed to income level.

The elasticities for food stamp recipients (prior to elimination of the purchase requirement) would be expected to be lower than that for eligible nonrecipients because of the provisions of the program. The Washington State data show this to be the case with an elasticity of .03 for the recipients.

The differences in elasticities between the Washington State data and the national BLS CES data can be attributed to differences in the sample. The state of Washington has relatively high Aid to Families with Dependent Children payments which places a relatively high floor under income for this particular sample of households with school-age children. The Washington State data include only households with 8- to 12-year-old children. The BLS CES data include households in all stages of the life cycle. A previous study using the 1955 USDA household budget data showed significant differences in income elasticities over the life cycle (Price 1969). However, elasticities were generally higher for the households with children than for those without, which suggests that the Washington State data should yield higher elasticities. In addition to sample differences, sampling and modeling error could cause the differences in parameter estimates.

Income elasticity estimates for the five major food groups from the 1973-74 BLS CES data were as follows: cereal and bakery products, 0.065; meat, poultry, and fish, 0.152; dairy products, 0.101; fruits, 0.195; and vegetables, 0.153.

The analysis of the consumption data on children yielded few significant relationships with current income. Four of the 20 items showed a significant relationship at the 0.10 level. The elasticity estimates for these were: beef, 0.08; all meat, 0.04; bread and rolls, -0.07; and cooked cereal, -0.10. Since the dependent variable was the quantity consumed and not expenditures, the lower elasticities would be expected.

An analysis of the frequency of the household serving major items was made using the Washington State data. Comparisons of frequencies of serving between households below and above 125 percent poverty were made. Of the 9 items analyzed, 4 (meat or fish, eggs, fruit, and potatoes) showed no significant relationship with poverty level. Dried vegetables and rice were served more frequently by the below poverty households while nondried vegetables, juice, and milk were served more frequently by the above poverty households.

A poverty level comparison was also made on the percentage of households usually serving specific items. Of the 159 items analyzed, 44 showed significant poverty level differences. Thus, the types of food usually served vary by poverty level but the differences are not extensive.

The evidence of the Washington State data shows small effects of income with respect to the quantities consumed by the child, the frequencies of serving major items, and the types of food served by the household. These results are not inconsistent with those found with the 1973-74 BLS CES expenditure data since income is expected to affect expenditures to a greater degree than quantities.

Assets. The Washington State study included data on both total and liquid assets. Total assets consisted primarily of the value of the home while savings accounts and checking account balances primarily constituted liquid assets. Correlation between liquid and total assets was so high that only one of these asset variables was included in any model of food expenditures or food consumption.

Crockett (1960) posits that the effect of assets on expenditures varies with the ratio of the household's actual to desired holdings of assets. Households whose actual levels of assets meet or exceed desired levels have a higher propensity to consume. The asset effect is hypothesized to be associated with life cycle. In some stages there is a desire to accumulate assets for such purposes as retirement, down payment on a home, or a college education. In other stages these needs do not exist. Since the Washington State data include only households with school-age children, life cycle stage should be relatively similar. The propensity to consume from assets should be relatively constant.

The model including all poverty levels showed that total assets significantly affect at-home food expenditures for the total sample and for the white subsample. Total assets were not significant for the black and chicano subsamples. For the total sample, a 1 percent increase in assets resulted in a 0.034 percent increase in food expenditures.

For the sample eligible for food stamps, the model included liquid assets. This variable was significant for the total eligible sample and for the subsample of eligible nonrecipients but not for the subsample of food stamp recipients. The latter results may be due to the provisions of the program. For the total eligible sample a 1 percent increase in liquid assets led to a 0.029 percent increase in food expenditures. For the eligible nonparticipants this percentage was 0.078.

Using the 1973-74 BLS CES data, the models included both homeownership and liquid assets. The homeownership variable was positive and significant for at-home expenditures but liquid assets showed no significant effects. At-home food expenditures for home owners were roughly 6 percent higher than for nonhome-owners. This data set showed home ownership to have a significantly negative impact on food-away-from-home expenditures.

Therefore, the effect of home ownership on total expenditures was negligible.

The 1973-74 BLS CES data for five major food groups showed that homeownership positively affects expenditures on three groups: cereal and bakery products, meat, poultry and fish, and dairy products. Expenditures by homeowners were 7.0, 8.3, and 7.5 percent higher, respectively, than nonhomeowners' expenditures. Expenditures on fruits and on vegetables were not significantly affected by homeownership.

The Washington State data showed that liquid assets affect consumption by the 8- to 12-year-old child. For 5 of the 20 food groups, consumption of the major dairy products, fresh fruits, pastries, and desserts was positively affected by liquid assets while consumption of fats and oils and cooked cereal was negatively affected. It should be noted that these effects were significant only at the 0.10 level.

The Washington State household data also showed liquid assets to positively affect the variety of fruits and vegetables usually served. From these data, 18 factors were formed from the types of fruit and vegetable items usually served. Liquid assets positively affected 8 of these factors. The factors affected were generally factors representing either fruits or fresh vegetables. The processed vegetable factors were not affected by liquid assets.

The Washington State data therefore indicate that the types and varieties of food served at home are affected by liquid assets. This data set also indicates that consumption of specific products and total at-home expenditures are affected as well.

Pay period. Length of pay period for the major income earner was included in the Washington State data. In the model for all households, food expenditures were 2.1 percent lower for the biweekly pay period than for the weekly period. There was some evidence that the effect of pay period was more pronounced on the low-income households. For those households eligible for food stamps, expenditures with biweekly and monthly pay periods were 6.9 percent and 9.7 percent, respectively, below that for the weekly period. Additionally, the effects were more pronounced among black and chicano households than among white households. Both assets and income on an adult equivalent basis were substantially lower for the chicano than for white households but relatively equal between the black and white samples.

Basic need levels. Of the five basic need levels (physiological, security, love and belonging, self-esteem, and self-actualization), physiological needs were included in the model explaining food expenditures for those households eligible for food stamps. Physiological needs represent the most fundamental of all human needs. The level of this need is individually determined and related to factors such as age, sex, and activity level. The degree of satisfaction of this need (and all other needs) depends on ability to attain goods and services (in this case, food) which meet the individual's level of physiological need. As this need becomes relatively satisfied, the motivation to attain food (and other goods related to physiological needs) declines and motivation for purchase of goods and services which satisfy higher needs intensifies. For a description of the instrument used to identify need level, see Price et al. (1980). The Washington State data showed food expenditures to be positively affected by the index of physiological need. The effect was significant at the 0.01 level for the sample of households eligible for food stamps and the subsample of food stamp recipients. The effect was not significant for the subsample of eligible nonrecipients. However, the coefficient was positive and nearly equal in size in the above sample and subsample.

     This finding has several implications. First, if the relationship between physiological need level and food expenditures is substantiated with other samples, models that do not include it will have biased estimates on some coefficients. According to Maslow, physiological need level is theoretically related to income level. Analysis of the Washington State data gave a degree of confirmation to the hypothesis. Without a measure of physiological need level in the model, the income elasticity estimates will, therefore, be biased.

     Second, the theory of basic need levels adds a degree of understanding to consumption behavior. It is a theory which has been used to help explain aspects of managerial behavior in different settings and can be used to understand varying aspects of consumption behavior. For example, these findings and the theory indicate that as income increases, basic physiological needs are satisfied and become less dominant as motivators of consumption behavior. The decline in physiological need would then decrease motivation to increase expenditures for food while at the same time increases in income would act to motivate increased food expenditures. This gives added insights into the reasons for low-income elasticities currently found for food expenditures.

     Need levels were also included in the models explaining the 8- to 12-year-old Washington State child's food consumption. The results showed that the consumption of various foods may respond to different need levels. Physiological need was positively related to the consumption of fats and oils. Since fats and oils are a very basic food, this finding is relatively consistent with the hypothesis. Consumption of mixed dishes was positively related to love and belonging needs. This result is also consistent with the theory since mixed dishes are frequently served at family gatherings. Mixed dishes would thus be associated with family gatherings and such gatherings satisfy love and belonging needs.

     Consumption of both fresh and canned vegetables were positively related to self-actualization need. This finding dovetails with the analysis of the number of fruit and vegetable items served by the household. Self-actualization need was positively related to both the number of fruits and the number of vegetables usually served. The child's consumption of pastries and desserts was also positively related to self-actualization need. These findings are consistent with the theory, since people with a high self-actualization need are creative and are concerned for all persons. They serve a large variety of foods. It would be a challenge to them to serve a variety of vegetables prepared in such a way as to be acceptable to members of their families, which should lead to increased vegetable consumption by the child. This same reasoning can be applied to pastries and desserts, even though these foods relate to very different nutrient needs.

The consumption of all meat was negatively related to all need levels except security needs. This finding does not appear to be consistent with the theory. It is either a spurious finding or the links with theory are not, at present, apparent.

Other relationships between the fruit and vegetable factors constructed from the Washington State household data and need levels were significant. Physiological need was negatively related to the fruit salad factor and to the fresh cabbage factor. This need was also negatively related to the number of salads and to the number of juices served. These foods may be perceived as not as essential to existence as other foods and, therefore, do not satisfy this need.

The love and belonging need was positively related to the fresh berry factor, the common canned vegetable factor, and to the number of fruits served. The common vegetable factor may be related to mixed dishes. That is, many of these vegetables may be included in mixed dishes. The high love and belonging need may lead to the serving of more fruits that are favorites of other members of the household.

In addition to the number of fruits and vegetables, the self-actualization need was positively related to the fresh berry and to the dried fruit factors. The desire for creativity and satisfying all persons in the household may lead to the serving of these relatively expensive fruit items.

FAMILY COMPOSITION, FOOD CONSUMPTION, AND QUALITY OF LIFE

This section illustrates the use of the Prais and Houthakker model in demand analysis.[8] Prais and Houthakker (1955) have proposed a behavioral model based on consumer equivalent scales. This approach provides a basis for analyzing the impact of selected socioeconomic characteristics on food purchases. It has been widely used in the literature (Nicholson 1949; Forsyth 1960; Price 1970; Singh and Nagar 1973; Kakwani 1977; Muellbauer 1979, 1980). Using an estimation method for the consumer unit scales developed by Chavas (1979), the project investigates family consumption behavior for 17 food items. It is shown that this approach can be particularly fruitful in analyzing the influence of selected socioeconomic factors on household food purchases and quality of life indices. Emphasis in this section is on some empirical results concerning the influence of family composition on food consumption and quality of life.

The data used in the analysis were obtained from the Consumer Expenditure Diary Survey conducted in 1972-73 by the Bureau of Labor Statistics. To conduct the analysis on groups as homogeneous as possible, only households with white and married family heads and located in the southern United States are considered. Furthermore, the analysis covers four different locations: (1) central cities, SMSA of more than 1 million population, (2) central cities, SMSA of 400,000 to 999,999 population, (3) central cities, SMSA of 50,000 to 399,999 population, (4) urban areas outside SMSA. The sample size is respectively 125, 67, 84, and 152 households for locations 1, 2, 3, and 4.

Family composition is characterized by the age structure of the household, that is, by the number of people (n_k) in different age categories. Five age categories have been selected: (1) 0-15 years old (n_1), (2) 16-25 years old (n_2), (3) 26-45 years old (n_3), (4) 46-65 years old (n_4), (5) over-65 years old (n_5).

An Engel relationship similar to that specified by Prais and Houthakker is used in this study to estimate the relationship for seventeen food items. It was first specified as a model linear in the variables. However, to take into account the fact that the marginal propensity to spend may vary with the income level and the family composition, interaction variables have been added to the model. Also, in order to capture the effects of location on food purchase behavior, the Engel parameters are assumed different for each location and a different Engel function is fitted for each location.

Since the earning power of a family is expected to vary with the age of its members (Ando and Modigliani 1963), then household income may be influenced by family composition. From this function, the effects of household composition on earning power can be estimated. These effects are expressed as earning power elasticities. An equation expressing income as a function of the life cycle variables and a dummy variable for college education is estimated. The equation has been specified as a model linear in the variables. It gives a representation of the life cycle hypothesis for income. The Engel function and the income function were estimated using ordinary least squares.[9]

Results

The analysis is conducted on weekly expenditures of 17 food categories and using weekly household income. The income equation is estimated for the four locations. The goodness-of-fit (R^2) varies from .51 to .75. College education of the family

head is found to have a larger impact on income in larger cities:
it increases weekly household income by $122.83, $112.75,
$105.49, and $55.54, respectively, for locations 1, 2, 3, and
4. As expected, the earning power is about zero for children,
increases with age up to 26–45 years of age, and then declines.
Also, the earning power of adults (26–45 years old) appears to
be larger in location 2 (cities between 400,000 and 999,999
population) than in other locations.

The Engel relationship is estimated for the 17 food items
and for the four locations. The average statistical fit (R^2) is
.25, which is reasonably good considering that the model is based
on cross-sectional data. From the estimates, income elasticities
are computed and presented in Table 10.4. Large variations appear across locations. For example, location 2 exhibits larger
income elasticities for beef and veal, poultry, eggs, dairy,
fruit, vegetables, and food away from home. Income elasticities
tend to be large for food away from home and negative for fats
and oils and alcoholic beverages. The income elasticity of total
food expenditures is 0.25, 0.56, 0 and 0.31, respectively, for
locations 1, 2, 3 and 4.

TABLE 10.4. Income elasticities for food items

Income elasticities	Location 1	2	3	4
Cereals	0.59	0.42	0.12	0.29
Beef and veal	0.19	1.78	0.13	−0.05
Pork	0.45	0.05	0.19	0.75
Other meats	0.96	−0.34	0.34	0.87
Poultry	−0.56	1.69	0.08	0.24
Fish and shellfish	0.31	0.63	−1.20	0.61
Eggs	−0.08	0.26	0.13	0.03
Dairy	−0.16	0.28	−0.09	0.05
Fruit	0.38	1.16	−0.12	0.80
Vegetables	−0.38	0.40	−0.22	0.07
Sugar and sweets	0.64	−0.06	0.45	0.55
Fats and oils	−0.42	−0.54	−0.76	0.43
Nonalcoholic beverages	0.11	0.09	0.17	0.34
Prepared food	−0.21	0.29	−0.36	−0.08
Food away from home	0.51	1.32	0.45	0.97
Alcoholic beverages	0.68	−0.38	−1.68	−0.53
Tobacco and smoking	0.26	−0.83	0.39	−1.01
Total food	0.25	0.56	0.00	0.31

Source: Chavas (1979).
Note: All elasticities are evaluated at the mean values of the
relevant variables.

The elasticities of food consumption with respect to family
composition are presented in Table 10.5 for location 2 (cities
between 400,000 and 999,999 population) and for the 17 food
items. The elasticities are evaluated at mean values. The
estimates show that cereals, other meats, poultry, eggs, dairy
products, vegetables, fats and oils, and nonalcoholic beverages
are typically consumed in greater amounts by children ($\leq$ 15

years old). Adults (26-45 years old) appear to consume relatively more beef and veal, fish and shellfish, fruits, sugar and
sweets, prepared foods, food away from home, and tobacco and
smoking supplies. Also, the introduction of young children in a
household tends to decrease the consumption of alcoholic beverages and smoking supplies. Similarly, the consumption of fats
and oils and alcoholic beverages declines as older people (> 65
years old) enter the family.

TABLE 10.5. Elasticities of consumption with respect to family
composition for location 2

Elasticities	0-15 years (n_1)	16-25 years (n_2)	26-45 years (n_3)	46-65 years (n_4)	Over 65 years (n_5)
Cereals	.29	.14	.23	.19	.07
Beef and veal	-.12	-.06	.37	.00	.05
Pork	.14	-.12	-.54	-.03	-.04
Other meats	.34	.17	.19	.22	.05
Poultry	.46	.22	-.28	.11	.06
Fish and shellfish	.05	.25	.31	.09	.05
Eggs	.42	.12	.04	.26	.03
Dairy	.25	.029	.19	.17	.07
Fruit	.08	.12	.32	.24	.07
Vegetables	.13	-.004	.00	.07	.02
Sugar and sweets	.27	.32	.36	.15	.07
Fats and oils	.19	.07	-.23	-.08	-.03
Nonalcoholic beverages	.22	.13	.17	.09	.02
Prepared food	.15	.23	.29	.18	.06
Food away from home	-.06	.14	.16	-.01	.03
Alcoholic beverages	-.19	-.14	-.02	.48	-.12
Tobacco and smoking	-.14	.31	.83	.20	.06

Source: Chavas (1979).
Note: All elasticities are evaluated at the mean values of the
relevant variables.

The elasticities of total food expenditures with respect to
family composition are presented in Table 10.6. In agreement
with results from previous research (Buse and Salathe 1979), the
addition of one adult (26-45 years old) to the family increases
food expenditures relatively more than the addition of any other
family member. Also, it is found that an increase in family
size always increases total food expenditures.

TABLE 10.6. Elasticities of food expenditures (E_{exp,n_k}) with respect
to family composition

Age category	Location 1	2	3	4
0-15 years (n_1)	.11	.06	.05	.08
16-25 years (n_2)	.10	.08	.09	.07
26-45 years (n_3)	.12	.15	.22	.11
46-65 years (n_4)	.06	.12	.13	.03
Over 65 years (n_5)	.09	.02	.05	.02

Source: Chavas (1979).
Note: All elasticities are evaluated at the mean values of the
relevant variables.

The elasticities of Table 10.6 are estimated from the Engel curve. In the context of the Prais and Houthakker model (1955) and using a procedure proposed by Chavas (1979), it is possible to solve for the elasticities of general scale and specific scales. Since all elasticities are estimated at some specified value (the mean values), the derived scale elasticities should be interpreted as valid only around these mean values. Alternatively, all elasticities reported give a local approximation to the true elasticities and should not be considered constant as income or family composition changes. This procedure has the advantage of simultaneously estimating general scale and specific scales that are consistent with each other and with the elasticities estimated from the Engel curve.

Following this approach, the elasticities of general scales can be calculated. As argued by Muellbauer (1974), these elasticities can be interpreted as cost-of-living indices in the absence of substitution effects. Results presented in Table 10.6 suggest that the introduction of one adult (16-45 years old) in the household increases the cost of living of a family by more than the introduction of any other family member. However, changing family composition also affects the earning power of households. The elasticities of income with respect to family composition are higher for adults and more so at location 3 (cities between 50,000 to 399,999 population) than at other locations. In this context, the net effect of a change in family composition can be measured by the difference between the earning power elasticity and the cost-of-living elasticity. This elasticity can be interpreted as an index of "quality of life" for different types of persons and different locations and used as a welfare evaluation criterion. Results in Table 10.6 suggest that location 3 cities have the highest quality of life index. Assuming that adults play a prominent role in family decision making, these results are in agreement with observed population migrations toward medium-sized cities. However, the indices vary across age categories. For example, children ($\leq$ 15 years old) and older people (> 65 years old) appear to be better off in large cities. These results appear reasonable and illustrate the usefulness of the approach to evaluate the welfare implications of selected sociodemographic variables.

CONCENTRATION CURVE ANALYSES OF FOOD CONSUMPTION

Blaylock and Smallwood (1982) examined disparities in the distribution of per capita food expenditures by income and urbanization in order to gain insight into the dietary and economic welfare of population subsets.[10] While the topic of inequality inherently occupies a more central platform in developing countries, it remains a germane issue in the United States where much government activity involves taxation, transfer payments, and social programs that tend to redistribute income. In addit-

ion to the traditional information that can be obtained by the
independent analysis of each distribution, it is possible to
simultaneously examine the expenditure and income distributions
within a structured economic framework to deduce the underlying
Engel (expenditure-income) relations. Other factors held constant, the distribution of expenditure on a commodity is directly related to the income distribution via the Engel function for
the commodity.

The project had three goals. The first was to investigate
urban and rural differences in the distribution and in the expenditure distributions for several aggregate food groups via
Lorenz and concentration curves and a related inequality measure.
The second purpose was to estimate and compare, in an interurban-
izational context, the behavior of expenditure (income) elasticities over the income spectrum. The elasticities are based
on a translated coordinate system for the Lorenz and concentration curves (Kakwani 1978). Elasticities derived using this
technique are not constrained to constant or monotonic relationships across the income spectrum and have the advantage of being
estimated using the prevailing income distribution. The final
objective was to introduce a generalization of the function form
originally proposed by Kakwani and Podder (1976) for the direct
estimation of Lorenz and concentration curves.

Lorenz and Concentration Curves

The Lorenz curve relates the cumulative proportion of total
income received to the cumulative proportion of population units
when the units are arranged in ascending magnitude of income.
Concentration curves relate the cumulative proportion of economic
variables other than income to the population distribution and
are generalized Lorenz curves. Both curves are directly related
via the functional relationship between income and other economic
variables of interest.

Equations of the Lorenz and concentration curves have been
traditionally derived from the density functions of the income
and expenditure distributions, respectively. Typically, the
curves have been fitted using some well-known density function,
usually with poor results (Kakwani and Podder 1976). An alternative approach (and the one employed) is to find functional forms
for the relationships which fit the data reasonably well. Given
estimates of the Lorenz and concentration curves, the corresponding expenditure elasticities and mean inequality coefficients
can be derived (Kakwani 1978).

Numerous functional forms are possible for describing the
relationships, but this paper will focus on a modified Beta
function and a generalized transformation of this function. The
Box-Cox transformation (Box and Cox 1964) is utilized to general-
ize and increase the flexibility of the model specification.
The Box-Cox functions allow the data more flexibility in determining the appropriate degree of nonlinearity rather than a
priori specification. The estimated functions are intrinsically

nonlinear in the parameters and were estimated by maximum likelihood techniques.

The first year of the Bureau of Labor Statistics 1973-74
Consumer Expenditure Survey (BLS CES) is the source of data for
this analysis. The sample is a nationally representative cross-
section of U.S. households that reports detailed information on
purchases during a two-week period together with socioeconomic
data concerning the household unit. Prior to this study, the
data was extensively examined for reporting and coding errors
and edited into a convenient format (Buse et al. 1978).

Annual income from all sources and average weekly expenditures for total food and eight major food groups were used for
the analysis of the food expenditure/income distribution relationship. The food groups examined included: (1) total food,
(2) food at home, (3) bakery products, (4) meat, poultry, fish,
and eggs, (5) dairy products, (6) fruits and vegetables, (7)
food away from home (FAFH), (8) cereal, and (9) other foods.
Income and food expenditures are placed on a per capita basis to
adjust for differences in family size.

The BLS CES sample was partitioned into three mutually exclusive and exhaustive urbanization categories. These three subgroups are defined as (1) SMSA--population of > 50,000, (2)
urban--population of 10,000-50,000, and (3) rural--population of
< 10,000. Separate analyses were performed on each level of
urbanization. The results of these analyses are presented in
the next section.

Results
     Both the modified Beta function and the Box-Cox specifica-
tion, as evaluated by several alternative criteria, are found to
be good choices for the functional form of the Lorenz and concentration curves. The values of R^2 are consistently high across
all food categories, income, and urbanizations indicating a good
overall representation of the actual data. A comparison of
actual and estimated mean income (expenditure) inequality coefficients reveals that both models do well in approximating the
functional relation at its maximum departure from the ordinate.
Additionally, actual and predicted income shares for the upper
and lower extremes of the distribution are very close for both
models. Thus, the overall model fit as well as the fit at the
extreme points is quite good.

The Box-Cox generalization of the modified Beta function is
a significant statistical improvement, as determined by the
likelihood ratio test, over the more easily estimated Beta
function for all cases considered. Although both models do
well, the Box-Cox formulation allows more flexibility in the
estimated distribution and more closely estimates the actual and
estimated mean income inequality coefficients in more than twice
as many instances. Perhaps more important for many applications, the Box-Cox model yielded more accurate results for the

predicted income and expenditure shares at the extremes of the distribution in every case estimated. In fact, the estimates become more accurate at the upper and lower 5 percent of the distribution than at the upper and lower 10 percent.

Food expenditures were found to be more equally distributed across the population than income. Thus, low-income families were allocating a larger share of their income for food than higher-income families. The expenditure and income inequality varied more between food groups than between urbanizations. The income inequality coefficient for all urbanizations was .30 compared to the expenditure inequality coefficient for total food of .12. This finding implies that the relative deviation of food expenditures from its mean is only 40 percent of that for income.

As expected, the largest expenditure inequality coefficient (EIC) obtained was for food away from home at .27. Food at home and its aggregate components had EICs which were considerably smaller. They were all found to be below .10. Cereal was found to have a small negative EIC, implying that expenditures decrease as income rises. The highest EICs for at-home food categories were obtained for fruit and vegetables, .073, and meat and eggs, .068.

The disparity in food expenditures was approximately equal across all urbanizations. The largest disparity was for food away from home (FAFH). The EIC for rural areas was estimated at .25 compared to .30 for urban (SMSA) areas. This result may partially reflect differences in convenience and accessibility of eating places in rural areas relative to more densely populated areas. Much less disparity across urbanizations was observed for food-at-home expenditure categories.

The elasticities of food group expenditures with respect to income were calculated using the estimated Lorenz and concentration curve parameters from the Box-Cox model. The elasticities were found to vary substantially across income classes, urbanizations, and food groups, and no simple functional relationship was able to characterize the behavior of the income elasticities over the income spectrum. In general, it was not found to be constant, linear, or monotonic across income classes. Elasticities evaluated at the median are close to those reported by Salathe (1978) using OLS estimation techniques with the same data base. Our results, however, indicate that the traditional linear (in parameters) specifications are not flexible enough to estimate elasticity relations across the income scale.

All income elasticities are positive over the income range with the exception of that of cereal products which is negative but close to zero. FAFH was the highest elasticity, ranging from 0.7 for the lowest incomes to 1.0 for the highest incomes. Income elasticities for food at home and its components have elasticities that are less than 0.5 across all income classes. The elasticity patterns exhibited by the various food groups are

more consistent across urbanizations than between food groups.
The patterns cannot be easily generalized.

SUMMARY

The empirical work reported in this chapter generally attempts to explain and predict food consumption and thereby identify variation among members of households with different socioeconomic profiles. The recognition of household characteristics provides insights as to how individual preferences are amalgamated into a household preference function (for food). This relaxes the assumption that the household is a closely knit decision unit whose food expenditure behavior may be explained primarily by traditional variables such as income. Emerging analytical techniques, translating, and scaling are discussed to show how socioeconomic characteristics can be incorporated into utility functions.

Other theoretical developments leading to introduction of psychological variables into food expenditure models have originated outside of traditional economic theory. Analyses from the Washington State data were unique in investigating the effect of basic need levels on types of food consumed and level of expenditures. This approach expands the dimensions of household characteristics affecting food consumption. Other variables such as assets that have not been included in many past models have received increased attention and have been identified as important factors that affect food consumption behavior.

Progress has been made towards identifying stable relationships. The methods of estimating coefficients for subgroups and analyzing differences in such coefficients discussed in this chapter are an initial step in identifying factors that cause changes in coefficients.

A summary of this chapter would not be complete without our best estimate of the overall income elasticity for major food groups in the United States. For a comparison of elasticities over time, three sets of income elasticity estimates for total food and for five major food groups by by Price (1969), Buse and Salathe (1979), and West and Price (1976) are presented in Table 10.7.

The elasticities generally decline over time. The exception is cereal and bakery products. Elasticities for these services

TABLE 10.7. Comparison of income elasticities over time

Commodity	1955 (Price)	1965 (Buse and Salathe)	1973–74 (West and Price)
Total food at home	.27	.23[a]	.10
Meats	.36	.30	.11
Cereal and bakery products	.10	.11	.09
Dairy products	.19	.15	.09
Vegetables	.28	.15	.14
Fruits	.71	.30	.15

[a]Income elasticity for total food in 1965 was estimated by Price to be .23 using the same methods as for the 1955 results.

are expected to be relatively high. Note the sharp decline in the size of the elasticities between 1965 and the 1973–74 period for meat and for total food. Sharp increases in the price of beef took place in 1973. Meat is a major component of total food expenditures and is a major source of the decline in that elasticity.

The decline in elasticities over time is an example of an empirical phenomenon which cannot be satisfactorily explained. Given the data in Table 10.7 and the knowledge incorporated in these estimates, it is difficult to predict the changes in the magnitude of the elasticities. Additional work on the differences in elasticities among groups of consumers and among different types of commodities, and the development of household decision-making theory are still required.

NOTES

1. Author of this section is Reuben C. Buse.
2. See Buse and Salathe (1979) for detailed estimates for the four household types for all 14 expenditure categories.
3. Buse and Salathe (1979) calculate similar tables for the other three types of households (i.e., homeowners in disequilibrium, and equilibrium and disequilibrium renters).
4. For details on how the socioeconomic and demographic variables were assumed to interact with income and the adult equivalent scale and how the marginal propensities and expenditure elasticities were calculated, see Buse and Salathe (1979).
5. Each expenditure elasticity is calculated by multiplying the estimated marginal propensity to spend by the corresponding average ratio between income and expenditure for households with that particular characteristic. These averages are available in Buse and Salathe (1979).
6. Authors of this section are David W. Price, Donald A. West, and Dorothy Z. Price.
7. A description of the Washington State sample of children can be found in Price (1977).
8. Author of this section is Jean-Paul Chavas.
9. The system is assumed recursive (i.e., the error term of the income equation is assumed uncorrelated with the error term of the consumption equations).
10. Authors of this section are   James Blaylock and David Smallwood.

REFERENCES

Agarwala, R., and J. Drinkwater. 1972. Consumption functions with shifting parameters due to socioeconomic factors. Rev. Econ. and Stat. 54:89–96.

Ando, A., and F. Modigliani. 1963. The life-cycle hypothesis of saving. Am. Econ. Rev. 53:55–84.

Barten, A. P. 1964. Family composition, prices, and expenditure patterns. Colston Pap. 16:277–92.

Benus, J., J. Kmenta, and H. Shapiro. 1976. The dynamics of household budget allocations of food expenditure. Rev. Econ. and Stat. 57:129–38.

Blaylock, J. R. and D. M. Smallwood. 1982. Engel analysis with Lorenz and concentration curves. Am. J. Agric. Econ. 64:134–39.

Bojer, H. 1977. The effects on consumption of household size and consumption. Eur. Econ. Rev. 9:169–93.

Box, G. E. P., and D. R. Cox. 1964. An analysis of transformations. J. R. Stat. Soc. Series B, 26:211–43.

Brandow, G. E. 1961. Interrelations among demands for farm products and implications for control of market supply. PA Agric. Exp. Stn. Bull. No. 680.

Brown, J. A. C., and A. Deaton. 1972. Surveys in applied economics: Models of consumer behavior. Econ. J. 82:1145–236.

Buse, R. C., J. S. Mann, and L. Salathe. 1978. Data problems in the 1972-1974 BLS Diary Survey Public Use Tapes. Univ. Wisconsin, Madison.

Buse, R. C., and L. E. Salathe. 1978. Adult equivalent scales: An alternative approach. Am. J. Agric. Econ. 60:460–68.

_____. 1979. Household expenditure patterns in the United States, 1960-1961; The last word. Univ. of Wisconsin Pap. 168, Madison.

Chang, O. H. 1979. Impact of permanent income, prices and sociodemographic characteristics on household expenditure patterns in the United States, 1960-61. Ph.D. diss., Univ. of Wisconsin-Madison.

Chavas, J. P. 1979. Consumer unit scales and food consumption. Pap., Dep. of Agric. Econ., Texas A and M Univ., College Station.

Cramer, J. S. 1971. Empirical econometrics. Amsterdam: North-Holland.

Crockett, J. 1960. Demand relationships for food. In Proceedings of the conference on consumption and saving, ed. I. Friend and R. Jones. Philadelphia: Univ. of Pennsylvania.

_____, and I. Friend. 1960. A complete set of consumer demand relationships. In Proceedings of the conference on consumption and saving, ed. I. Friend, and R. Jones, Philadelphia: Univ. of Pennsylvania.

Engel, E. 1895. Die lebenskosten belgischer arbeiter-familiar and yetzt. Int. Stat. Inst., Bull. g, Rome:1–124.

Ferber, R. 1962. Research on household behavior. Am. Econ. Rev. 54:19–63.

_____. 1973. Consumer economics, a survey. J. Econ. Lit. 11:1303–342.

Forsyth, F. G. 1960. The relationship between family size and family expenditure. J. R. Stat. Soc. Series A, 123:367–97.

Friedman, M. 1952. A method of comparing incomes of families of differing compositions. Stud. Income and Wealth 15:9-74.
_____. 1957. A theory of the consumption function. Princeton: National Bureau of Economic Research.
George, P. S., and G. A. King. 1971. Consumer demand for food commodities in the U.S. with projections for 1980. Giannini Found. Monogr. No. 26, Univ. of California, Berkeley.
Goldstein, K. 1939. The organism. NY: American Book Co.
Goldstein, S. 1966. Urban and rural differentials in consumer patterns of the aged, 1960-61. Rur. Sociol. 31:353-54.
Green, R., Z. A. Hussan, and S. R. Johnson. 1978. Alternative estimates of static and dynamic demand systems for Canada. Am. J. Agric. Econ. 60:93-108.
Hamburg, M. 1960. Demand for clothing. In Proceedings of the conference on consumption and saving, ed. I. Friend and R. Jones. Philadelphia: Univ. of Pennsylvania.
Hassan, Z. A., and S. R. Johnson. 1976. Family expenditure patterns in Canada: A statistical analysis of structural homogenity. Agric. Can. (Ottawa), Econ. Br. Publ. No. 76/3.
_____. 1977. Urban food consumption patterns in Canada. (Ottawa), Econ. Br. Publ. No. 77/1.
Kakwani, N. C. 1977. On the estimation of consumer unit scales. Rev. Econ. and Stat. 59:507-10.
_____. 1978. A new method of estimating Engel elasticities. J. Econometrics 8:103-10.
_____, and N. Podder. 1976. Efficient estimation of the Lorenz curve and associated inequality measures from grouped observations. Econometrica 44:137-48.
Kapteyn, A., and B. Van Praag. 1976. A new approach to the construction of family equivalent scales. Eur. Econ. Rev. 7:313-35.
Lai, L. K. 1972. The estimation of effects on expected family income and socioeconomic variables of U.S. household consumption of food. Ph.D. diss., Univ. of Wisconsin-Madison.
Lansing, J. B., T. Lorimer, and C. Moriguchi. 1960. How people pay for college. Ann Arbor: Survey Research Center, Univ. of Michigan.
Lau, L. J., W. L. Lin, and P. A. Yotopoulos. 1978. The linear logarithmic expenditure system: An application to consumption-leisure choice. Econometrica 46:843-68.
Lee, F. Y., and K. E. Phillips. 1971. Differences in consumption patterns of farm and nonfarm households in the United States. Am. J. Agric. Econ. 53:573-82.
MacMillan, J. A., F. L. Chung, and R. M. A. Loyns. 1972. Differences in regional household consumption patterns by urbanization: A cross-sectional analysis. J. of Reg. Sci. 22:417-24.
Maisel, S. J., and L. Winnich. 1959. Family housing expenditures: Elusive laws and intrusive variances. In Proceedings of the Conference on consumption and saving, ed.

I.   Friend   and   R.   Jones.   Philadelphia: Univ.   of
        Pennsylvania.
Maslow, A.   1970.   Motivation and   personality.   NY: Harper   and
        Row.
McClements,   J. D.   1977.   Equivalent   scales   for   children.   J.
        Public Econ.   8:191-210.
Morgan, J. N.   1965.   Housing and   ability to pay.   Econometrica
        33:289-306.
Muellbauer, J.   1974.   Household composition,   Engel curves,   and
        welfare   comparisons   between   households: A   quality
        approach.   Eur.   Econ.   Rev.   5:103-22.
_____.   1975.   Identification and consumer   unit scales.   Econo-
        metrica 43:807-9.
_____.   1977.   Testing the Barten model of household composition
        effects and the cost of children.   Econ.   J.   87:460-87.
_____.   1979.   McClements   on   equivalence scales for children.
        J.   Public Econ.   12:221-31.
_____.   1980.   The estimation of the   Prais-Houthakker   model of
        equivalence scales.   Econometrica 48:153-76.
Nicholson, J. L.   1949.   Variations in   working   class family ex-
        penditure.   J.   R.   Stat.   Soc.   Series A,   112:359-411.
Orshansky, M.   1965.   Counting   the   poor,   another   look at   the
        poverty profile.   Soc.   Secur.   Bull.   28:3-29.
Parks, R. W.,   and   A. P. Barten.   1973.   A cross-country   compar-
        ison   of   the   effects   of   prices,   income,   and   population
        composition on consumption patterns.   Econ.   J.   83:834-52.
Phlips, L.   1974.   Applied   consumption   analysis.   Amsterdam:
        North-Holland.
Pollak, R. A.,   and   T. J. Wales.   1978.   Estimation   of   complete
        demand systems from household budget data: The linear and
        quadratic expenditure systems.   Am.   Econ.   Rev.   68:348-59.
_____.   1979.   Welfare comparisons and equivalence scales.   Am.
        Econ.   Rev.   69:216-22.
_____.   1980.   Comparison of the   quadratic   expenditure   system
        and translog demand systems with alternative specifications
        of demographic effects.   Econometrica 48:595-612.
Prais, S. J.,   and   H. S.   Houthakker.   1955.   The analysis   of
        family budgets.   Cambridge: Cambridge Univ.   Press.
Price, D. W.   1969.   The   effects   of   household   consumption   on
        income elasticities   of   food   commodities.   Washington State
        Univ.   Tech.   Bull.   No.   63.
_____.   1970.   Unit equivalent scales for specific food commodi-
        ties.   Am.   J.   Agric.   Econ.   52:224-33.
_____.   1977.   Food patterns   in Washington households   with   8-12
        year   old   children.   Washington State Univ.,   CARC Bull.   No.
        843.
_____ et al.   1976.   Evaluation   of   school   lunch   and   school
        breakfast programs in the state of Washington.   Final rep.
        submitted to Food and Nutr.   Serv.,   USDA.
Price, D. W.,   D. Z. Price,   and D. A. West.   1980.   Traditional
        and non-traditional determinants of household expenditures

on selected fruits and vegetables. West. J. Agric. Econ. 5:21–36.

Salathe, L. E. 1978. A comparison of alternative functional forms for estimating household Engel curves. Pap. at meetings of Am. Assoc. Agric. Econ., Virginia Polytechnic Inst. and State Univ.

_____, and R. C. Buse. 1979. Household food consumption patterns in the U.S. USDA Tech. Bull. No. 1587.

Samuelson, P. A. 1966. Foundations of economic analysis. Cambridge, MA: Harvard Univ. Press.

Seneca, J. J., and M. K. Taussig. 1971. Family equivalence scales and personal income tax exemptions for children. Rev. Econ. and Stat. 53:253–62.

Sherwood, M. 1975. Family budgets and geographic differences in price levels. Mon. Labor Rev. 98, No. 4:8–15.

Singh, B., and A. L. Nagar. 1973. Determination of consumer unit scales. Econometrica 41:347–55.

Sobel, M. 1960. Correlates of present and future work status of women. Doctoral diss., Univ. of Michigan.

Sonquist, J. A., E. L. Baker, and J. M. Morgan. 1973. Searching for structure. Surv. Res. Cent., Inst. for Soc. Res., Univ. of Michigan.

Sydenstricker, E., and W. I. King. 1921. The measurement of relative economic status of families. Q. Pub. Am. Stat. Assoc., NS, 17:842–57.

U.S. Department of Agriculture. 1972. Food consumption of households in the United States and year, 1965–66. Rep. No. 12.

Watts, H. 1958. Long run income expectations and consumer savings. Cowles Found. Discuss. Pap. No. 123, New Haven, CT.

West, D. A. 1979. Effects of food stamp program on food expenditures: An analysis of the BLS consumer expenditure survey 1973–74 diary data. Rep. to Food and Nutr. Serv., USDA.

_____. 1979. Food expenditures and the Food Stamp Program: Some recent evidence from the consumer expenditure survey. Pap., Washington State Univ.

_____, and D. W. Price. 1976. The effects of income, assets, food programs, and household size on food consumption. Am. J. Agric. Econ. 58:725–30.

West, D. A., D. W. Price, and D. Z. Price. 1978. Impacts of the food stamp program on value of food consumed and nutrient intake among Washington households with 8–12 year old children. West. J. Agric. Econ. 3:131–44.

PART IV

Policy Issues
Affecting Nutrition

CHAPTER 11

Consumer Demand
for Nutrients in Food

CONSUMERS purchase goods because of the utility they provide. The utility of goods arises from the characteristics or attributes that the goods possess. Economic theory holds that <u>ceteris paribus</u> the marginal utility of a product is reflected in price; furthermore, that the market conditions under which a product is purchased also impact on price. It remains for empirical analysis to determine both the relative weights given by consumers to each product characteristic as reflected by price and the influence of market factors on prices. It is to these issues that this chapter is addressed.

The decision process which results in the purchase of a particular commodity at a given price is obviously quite complex. Some conceptualization of the process is thus useful in understanding consumer behavior and the observable factors which influence it. The decision process is conventionally described in terms of objectives or values to be fulfilled, some facts about alternatives, and some projected outcomes of decisions drawn from the objectives and facts. Brim et al. (1962) identified six sequential phases in the decision-making process in a problem solving framework. These phases included identification of the problem, obtaining necessary information, development of alternative possible solutions, evaluation of the alternatives, selection of a strategy for performance or course of action, and actual performance. Concurrent with the terminating action for one process they identified subsequent learning and revision of strategy. Thus they linked one process with another.

In the case of making a decision about which of two products to buy, identification of the problem would refer to awareness of wants for particular attributes which the goods in question might supply. To obtain information necessary for making the

---

Author of this chapter is Karen J. Morgan.

decision, buyers discuss needs with their families and seek information from advertisements, product labels, comparison shopping expeditions, their friends, buying guides, technical specialists, and other sources. As buyers acquire information, they develop ideas regarding alternative possibilities and revise perspectives about the criteria which are pertinent to making final choices among alternatives. When a buyer has decided what to buy, the decision of where and when to buy and within what price range has to be made. The actual purchase is the event which terminates the process. However, the process might also be terminated by a decision not to buy at that time any product having the attributes which were being sought. If the buyer makes the purchase, an expression of the buyer's evaluation of the bundle of attributes inherent in the marginal unit of the good is expressed through the price mechanism.

Increased public interest in nutrition has heightened the value of economists obtaining a better understanding of the purchasing behavior of buyers when obtaining food items in the marketplace. Waugh (1927) concluded that there was a distinct tendency for market prices of food to vary with certain physical characteristics which the consumer identifies with quality. However, until recently Waugh's findings have had little impact on the theory and measurement of consumer demand. All intrinsic properties of particular goods, those properties that make a mink coat different from a quart of milk, have been omitted from the traditional theory of consumer economics. The results of traditional demand theory are highly dependent on lack of product disaggregation. Consequently, the only property which this theory can build on is the property shared by all goods, which is simply that they are goods.

Recognizing the unsuitability of traditional theory for dealing with consumer demand, Lancaster (1966) broke away from the traditional approach of treating goods as the undifferent-iated totality of utility, and instead, proposed that it is the properties or characteristics of the goods from which utility is derived. Lancaster assumed that consumption is an activity in which goods, singly or in combination, are inputs and in which the output is a collection of characteristics. Utility or pre-ference orderings rank collections of characteristics and only rank collections of goods indirectly through the characteristics that they possess. Lancaster argued that, in general, a good possesses more than one characteristic, and many characteristics are shared by more than one good; furthermore, goods in combina-tion may possess characteristics different from those pertaining to the goods separately. Lancaster assumed the characteristics possessed by a good or a combination of goods are the same for all consumers, and that they are measured in the same quantity units; thus the personal element in consumer choice arises in the choice between collections of characteristics, not in the allocation of characteristics to goods. Therefore, the objective nature of the goods-characteristics relationship plays a crucial role in Lancastern theory.

Ladd and Suvannunt (1976) and Ladd and Martin (1976) have expanded on the idea that products are wanted because of the utilities they provide and that the utilities provided depend upon the product characteristics. Like Lancaster, they maintain that the total amount of utility a consumer enjoys from the purchase of products depends upon the total amounts of product characteristics purchased. Ladd and Suvannunt have developed a model--the consumer goods characteristics model (CGCM)--that yields two hypotheses: (1) for each product consumed, the price paid by the consumer equals the sum of the marginal monetary values of the product's characteristics, and the marginal monetary value of each characteristic equals the quantity of the characteristics obtained from the marginal unit of the product consumed multiplied by the marginal implicit price of the characteristic; and (2) the consumer demand functions for goods are affected by characteristics of the goods.

An important underlying assumption of the CGCM is that consumers can decide how much of a product to buy but they cannot decide the amount of each characteristic to be contained in or provided by one unit of a product. Assuming consumers can transmit their desires to producers, the researchers in the hedonic price index area (e.g., Adelman and Griliches 1961; Dhrymes 1967; Griliches 1968) base their theory on the premise that consumers' purchasing power influences not only the amount of product produced but also the magnitude of the characteristics contained therein.

The essence of the hedonic technique is the disaggregation of products into characteristics and the estimation of implicit prices for units of the characteristics. Thus, in this respect, the hedonic technique is quite similar to Ladd and Suvannunt's CGCM. However, the hedonic technique goes further by using implicit prices of characteristics of goods for adjusting the observed market prices for different product varieties for the value of quality differences. The theoretical base of such studies was specified by Adelman and Griliches.

A reasonably large literature on the subject as it relates to durable goods has been developed (Dean and DePodwin 1961; Bailey et al. 1963; Fetting 1963; Triplett 1963; Brown 1964; Cagan 1965; Dhrymes 1967; Fisher et al. 1967; Gavett 1967; Griliches 1968; Kravis and Lipsey 1969; Musgrave 1969; Gillingham and Lund 1970).

Although most hedonic investigations have centered on the study of durable goods, there have been a limited number of studies which have estimated relations between prices of food and their characteristics. Waugh (1927) collected data on wholesale prices and characteristics of individual lots of asparagus, tomatoes, and cucumbers. He regressed measures of product characteristics on the ratio of the price of each lot to the average price of the product and converted the regression coefficients into prices of product characteristics. Ladd and Suvannunt used a linear form equation to relate annual average retail prices of 31 different meat, dairy, and poultry items to amounts of various nutritional elements they provided per pound.

The investigations by Waugh 1927 and by Ladd and Suvannunt 1976 have not dealt adequately with the implications of their findings, especially the negative coefficients obtained for some of the product characteristics. This study further resolves some of the problems involved in developing a hedonic index for nondurable goods.

Results presented here, which are an extension of previously published results (Morgan et al. 1979), are useful in better understanding the effects of nutrient composition of food on purchasing behavior, as well as in highlighting some of the problems associated with using the commodity characteristics approach in studying food purchasing. Breakfast cereals were selected for study because they are a labeled, regularly purchased food product and contain nutrient supplements. The product is thus one for which conditions are favorable for obtaining results that may demonstrate the potential this approach has in the study of food purchases.

MODEL SPECIFICATION

Nineteen dietary components, five other product characteristics, and four market factors were selected as the independent variables to be estimated for the hedonic index.[1] Although the market factors are not breakfast cereal characteristics, they are included to condition price, that is, they are variables used to statistically control for differences in circumstances under which data were developed.

Pretesting of several model specifications showed all qualitative factors (e.g., other product characteristics and market factors) to have reasonable price coefficients.[2] However, the same testing revealed that the estimated implicit prices were erratic and inconsistent with a priori information for the dietary variables. Thus, all dietary variables were omitted from the regression analysis in an alternative model specification to determine the importance of their incorporation. A statistical evaluation of this restricted model indicated that the set of dietary variables had a significant impact on the price. Even though unexpected signs and magnitudes of coefficients were obtained for some of the individual dietary variables, these variables, as a group, had a systematic influence on cereal prices. Thus, it appeared that a better model could be developed by altering the way in which the dietary variables were specified.

A logical approach to respecifying the hedonic index model resulted from research previously reported by Cohen et al.

(1972) and Mazis et al. (1975). The Cohen et al. multiattribute
model, labeled "adequacy-importance" model, assesses an
individual's attitude toward an object on the basis of (1) how
important the attributes of that object are to the individual
and (2) the evaluation of the object with respect to those
attributes. Mazis et al. pointed out that attributes are
sometimes defined as "product dimensions, rather than specific
characteristics to be used (in the model)." Since the unrestricted models yielded unexpected signs and magnitudes of
coefficients for some of the dietary component variables, it was
decided that these very specific characteristics were not of
value or were not evaluated by buyers. A researcher could
conclude on this basis that the dietary components of breakfast
cereals are not relevant attributes to buyers. However, if
producers are rational in adding costly supplements to cereals,
then these supplements must be of value to consumers and thus
must have positive implicit prices. Thus, in this particular
hedonic index formulation Cohen et al.'s definition of attributes
was heeded; that is, product dimensions, having multiple characteristics rather than specific characteristics were used, at
least partially, in specifying a hedonic index for breakfast
cereals.

It was assumed that buyers assess the dietary components of
breakfast cereals (at least to a limited extent). However,
evidently the buyers do their assessment procedure in a conglo-
merate way: buyers choose cereals which contain numerous
vitamins and minerals but they do not know or precisely evaluate
each of those contained within a particular cereal. It was
postulated that buyers prefer more to fewer vitamins and minerals
in cereals. It was also assumed that buyers prefer an even
distribution of these vitamins and minerals, that is, buyers
prefer that cereals be supplemented with vitamins and minerals
at similar percentage of U.S. Recommended Daily Allowance (RDA).
In addition to vitamins and minerals, it was postulated that
buyers are aware of protein and caloric content of cereals and,
to a lesser extent, fiber content.

A number of approaches might be taken in justifying the
restricted specification of the model. For example, investigations have indicated that individuals tend to evaluate foods on
the basis of key dietary factors or view nutritional components
of foods as within major nutrient groups (Federal Trade Commission 1976). Economically, the specification might be justified
on the basis of the information cost to the buyer of obtaining
the background for completely evaluating the dietary components
(Schwartz 1975; United States Department of Agriculture 1975).
Whatever the basis for the more restricted specification, the
assumption is that buyers are processing the detailed dietary
information available on the cereal label in an altered form.

With the assumptions on grouping of dietary variables, the
model for estimating the coefficients for the hedonic index was
specified as

$$P_{it} = \sum_{j=1}^{n} \beta_{jt} X_{jit} + \sum_{Y=1}^{m} \delta_{rt} Z_{rit} + \mu_{it}$$

where P_{it} is the cereal price per ounce, X_{jit} represents the dietary characteristics, Z_{rit} denotes the nondietary characteristics and market factors, μ_{it} is the structural disturbance, and i and t indicate the cereal and time period, respectively. The parameters β_{jt} and δ_{rt} are to be estimated for forming the hedonic index.

DATA AND ESTIMATION PROCEDURES

Price data for this investigation were obtained by requesting information on cereal purchases from state purchasing agents for all 50 of the United States, for the period 1972 through 1976. Contract information (vendor, product, size of package, price, basis of price quotation, and length of contract) was received from 12 states.[3] State purchasing agents were assumed to reflect the preferences of their clientele.

There are differences in what one would expect from data on state purchasing agents and similar information on individual consumers (state purchasing agents presumably may have greater knowledge and market power). However, since this was an exploratory study on utilization of the specified model and since state purchasing agents and individual consumers are similar in that both strive to minimize expenditures to achieve a specified level of satisfaction, the implication that the purchasing behaviors between the two types of buyers might be different was not of particular concern to the research. When making selections of what food items to buy, it can be argued that purchasing agents are influenced by their clients' preferences. For example, if their clients do not consume the cereals purchased, these cereals will not be purchased again. Thus, to some extent, the decisions of purchasing agents reflect the preferences and purchasing inclinations of individual consumers. The major limitation of generalizing from state purchasing agents to consumer behavior is that the clientele (consumers) do not make consumption decisions on a particular cereal based on its price. For example, when a client selects a box of cereal from a cafeteria line, she or he does not consider the price of each individual cereal as they are all priced the same in such a situation. This limitation of the study is recognized and considered in concluding remarks. Relevant manufacturers were contacted to obtain information on the dietary components and other pertinent product characteristics for each individual breakfast cereal.

The price data were separated by contract period (1972-73, 1973-74, 1974, 1974-75, 1975, 1975-76, and 1976). The hedonic

index was constructed so that implicit prices per unit of specified characteristics reflected a situation in which each
variety of the product contributed approximately equally to the
formulation of the index. Although these data reflect amounts
purchased in given time periods, they result in an unequal
representation of vendors, products, and market factors (i.e.,
state where purchased, basis of price quotation, length of
contract, and size of package). For more uniform representation, the data for each time period were simultaneously sorted
according to these six factors. The mean price was then obtained for each cereal purchased under identical conditions for each
given time period (i.e., same vendor, product, basis of price
quotation, length of contract, package size, and state where
purchased). These means were used as the price observations for
the statistical analysis.[4] Thus, theoretically, there is one
observation in each time period for each combination of these
variables. Although the distribution of observations still did
not reflect complete equality, it was at least approximated.
The resulting sample size was 1,702 with the observations distributed across contract periods, cereal classes, and other characteristics in a fairly representative manner.[5]

Vitamin and mineral contents were presented by a mean percentage of U.S. RDA for all vitamins and minerals provided by an
ounce of each cereal. These values were obtained by summing the
reported percentages of U.S. RDA for all vitamins and minerals
contained in an ounce of a given cereal and dividing by the
highest potential number of vitamins (10) and minerals (6) found
in cereals. Vitamin and mineral dispersion values were obtained
by subtracting the percentage of U.S. RDA for each vitamin or
mineral from the mean percentage of all vitamins or minerals
combined, squaring the difference, summing the squared values,
and dividing the total by the number of vitamins or minerals
minus one, (i.e., 9 or 5). Thus, both variance variables were
computed by standard methods. Protein and fiber were expressed
in grams per ounce, and calories in hundreds of calories per
ounce. All other independent variables except length of contract
(expressed in months) were qualitative and, therefore, had values
of 0 or 1.

In addition to its application to the data for all cereals,
the hedonic model was also used to analyze the data categorized
by class of cereal (high-nutritive, "kids' stuff," traditional).[6]
This categorization was done because previous experimentation
indicated that different weights were given to the characteristics depending on cereal class.

Ordinary least squares was used to estimate the parameters
for each contract period. The difference in parameters among
the periods did not prove to be statistically significant, so
the data were pooled. Calculation of the variances of the
disturbances showed them to be similar for the periods. Thus,

the variance for it was assumed identical for the periods and ordinary least squares was applied to the pooled data for obtaining the results presented.

RESULTS AND DISCUSSION

Analysis of All Cereals for Dietary Components

The estimated coefficients resulting from analysis of all cereals combined are presented in Table 11.1. Protein, vitamins, minerals, and vitamin variance were found to be positively related to price. The coefficient for protein was found to be statistically significantly different from zero at the .001 level of probability, and the coefficient for minerals, statistically significant at the .05 level of probability. A negative relationship between price and mineral variance, fiber, and calories indicate that breakfast cereal buyers, in general, consider these attributes to be less important characteristics for cereal products. There was a positive relationship between vitamin variance and price, although the coefficient had a probability level of only .191. This implies that buyers prefer cereals which contain a large percentage of U.S. RDA of particular vitamins and smaller percentages of U.S. RDA for other vitamins. The negative sign for the statistically significant mineral variance coefficient implies that buyers prefer to have a wide, even distribution of minerals in their cereals (i.e., to obtain a high price for mineral supplementation companies need to provide an even distribution of minerals in their cereals).

These opposite signs for the two variance variables are understandable in light of general consumer knowledge. Many adults are aware of at least some of the vitamins that need to be consumed to maintain good health; they seem to be particularly aware of the need for specific ones such as vitamin C and vitamin A (Federal Trade Commission 1976). On the other hand, considerably fewer people are aware of the body's need for specific minerals. Many people have heard that the human body needs minerals in order to function properly; however, just which minerals are needed is not known by many individuals (Schwartz 1975). Thus, buyers in general tend to select breakfast cereals which contain the specific vitamins they have heard (and probably believe) are essential for good health, whereas for decisions pertaining to mineral content, individuals select a cereal which contains many minerals which have a fairly even distribution in cereals since they know less about mineral needs.

The negative sign for the coefficient for calories is understandable in light of the American awareness of weight control. Nearly all individuals are aware of caloric content of food; furthermore, the majority of people in America are interested in consuming foodstuff that is low in calories (Federal Trade Commission 1976). The negative coefficient for calories is consistent with this belief, that is, as caloric content increases, implicit value of the cereal decreases. In light of the recent

TABLE 11.1. Estimated coefficients for characteristics and market factors of breakfast cereals

Characteristic and market factor	All cereals combined	High-nutritive	"Kids' stuff"	Traditional
Dietary component				
Protein	.008[c]	.009[c]	−.004[b]	.005[c]
Vitamins	.010	.135[a]	.066	.069
Minerals	.067[a]	.281	.063	.184[c]
Vitamin variance	.087	−.267	.256	−.473
Mineral variance	−.165[b]	−.469[b]	−.157	−.503[c]
Fiber	−.006[b]	−.068	−.018	−.008[a]
Calories	−.014[b]	.156	−.071[a]	.010
Package size[d]				
Individual serving	.016[c]	.020[c]	.013[c]	.016[c]
Preparation required[d]				
Ready-to-eat	.069[c]	--	--	.086[c]
Class of cereal[d]				
"Kids' stuff"	−.001	--	--	--
Traditional	.003	--	--	--
Type of processing[d]				
Flaked	−.029[c]	--	−.022[c]	−.033[c]
Shredded	−.027[c]	--	−.018[c]	−.025[c]
Granulated	−.034[c]	--	--	−.035[c]
Oven-toasted	−.034[c]	--	--	−.031[c]
Milled	--	--	--	--
Price quotation[d]				
Winning bid	−.003[b]	−.019	−.001	−.003[a]
Negotiated	−.011[c]	−.012	−.006	−.018[c]
Production company[d]				
General Foods	−.004	--	−.019[a]	−.003
General Mills	−.005	--	−.020[a]	−.009
Kellogg	−.001	--	−.017	−.002
Quaker Oats	.025[c]	--	−.018	.043[c]
Ralston Purina	.005	--	−.009	.008
Standard Milling	.023[c]	--	--	.029[c]
State of purchase[d]				
Arkansas	−.003	.006	.001	−.003
Florida	−.008[c]	−.001	−.004	−.009[b]
Hawaii	.017[c]	.028[c]	.025[c]	−.012[c]
Maine	−.029[c]	−.027	--	−.037[c]
Massachusetts	−.014[c]	--	--	−.024[c]
Minnesota	−.014[c]	--	−.010[a]	−.022[c]
Missouri	−.011[b]	−.007	−.006	−.019[c]
Nebraska	−.003	.002	−.003	−.004
North Dakota	−.001[c]	−.001	−.004	−.009[c]
Oklahoma	−.013[c]	--	−.007	−.022[c]
Virginia	−.002	−.002	−.002	−.001
Principal grain[d]				
Corn	−.002	--	−.008[c]	.001
Wheat	−.005[b]	.046[a]	−.022[c]	.006[a]
Oats	−.018[c]	−.003	−.010[c]	−.007[a]
Rice	.021[c]	--	−.003	.022[c]
Barley	−.008	--	--	−.005
Length of contract	.000	.000	.000	.000
R^2	.731	.845	.706	.792

Source: Morgan, et al. (1979), with permission.
Note: Dash (--) signifies that no coefficient could be estimated.
[a]Significantly different from zero at .05
[b]Significantly different from zero at .01
[c]Significantly different from zero at .001
[d]Variables appearing in the intercept include bulk package, needs-to-be-cooked, high-nutritive, puffed, losing bid, Van Brode, Washington, mixed grains.

publicity informing people that their bodies need fiber and thus they should consume foods with a high fiber content, one might have expected fiber content to be positively related to price. However, the statistically significant (P $\leq$ .01) negative relationship between price and fiber may be rationalized in light of the time period this investigation covered. People have not, at least generally, been made aware of the human body's need for fiber until recently, probably during the past five years. These data are from 1972 through 1976, and thus the majority of the data covers a time period during which there was limited awareness of the body's need for fiber.

Analysis of Three Classes of Cereals for Dietary Components

Initial analyses of the data indicated the value of partitioning the sample by class of cereal and analyzing the classes separately, as there appeared to be different weights given to the characteristics of breakfast cereals depending on which class of cereal was purchased.

The signs and estimated coefficients indicate that buyers who purchase high-nutritive cereals and those who purchase traditional cereals have similar purchasing behavior (Table 11.1). Both types of buyers value high protein content in breakfast cereals. They prefer cereals which are supplemented with many vitamins and minerals and which are supplemented in such a way that these vitamins and minerals are fairly evenly distributed. Somewhat suprisingly, these two types of buyers appear to prefer cereals which contain an above average amount of calories and a lower average amount of fiber.

When findings for traditional and high-nutritive cereals are compared with those for "kids' stuff" cereals, some diverse results are evident. "Kids' stuff" cereal buyers appear willing to pay additional money for vitamin and mineral supplementation but not for increased protein levels. These buyers apparently look for cereals which have an even distribution of minerals, but select cereals which contain a supplementation of specific vitamins. It appears that these buyers also have little interest in cereals which are high in calories or fiber.

Analysis of Qualitative Factors

The coefficients for the qualitative factors, which are interpreted as price differentials, may be observed in Table 11.1. Some implicit prices were not generated, due either to nonexistence of a factor (e.g., no "kids' stuff" cereals are of the needs-to-be-cooked variety) or to lack of variation within sample (e.g., type of processing for high-nutritive cereals).

The effect of package size showed that individual serving packages contributed more to price than bulk form. Results from analysis of all cereals combined and the traditional class of cereal indicate that ready-to-eat cereals commanded higher prices than did needs-to-be-cooked cereals. The effect of type of processing showed the puffing process to have a greater impact

on price per ounce than any other process. Assessment of the
implicit prices for principal grain showed that for all cereals
combined and for the traditional class, rice had higher price
coefficients than all other grains; whereas for "kids' stuff"
cereals, mixed grains had the highest price coefficient. In the
case of high-nutritive cereals, wheat was found to contribute
more to price than oats and mixed grains.

Assessment of market factors indicates that losing bids
consistently contributed more to price quotation than did the
other two bases for price quotations. In all groupings, Hawaii
had the highest price coefficients and Virginia, the lowest.
The inconsistent results for manufacturers among the four groupings of cereal indicate that pricing patterns of companies for
different classes of cereals are probably not consistent. The
information obtained on manufacturers for all cereals combined
cannot lead to any precise conclusion, as all companies do not
produce all types of cereals.

Interpretation of Estimated Coefficients

The coefficients, levels of probability for the coefficients, means and standard deviations of the dietary components,
and contribution of the dietary components to the price of one
ounce of various types of cereal are displayed in Table 11.2.
When all cereals were combined most of the coefficients appeared
to be reliable estimates for the dietary components of breakfast
cereals, that is, the levels of probability were within a toler-
able range. However, the reliability of the coefficient for
vitamins is somewhat questionable since the probability level
for it is .557.

More of the estimated coefficients for high-nutritive cereals
appear to be unreliable due to their high level of probability
(e.g., minerals, vitamin variance, calories, and fiber had probability levels of .236, .265, .545, and .208, respectively).
The "kids' stuff" cereal data also revealed some coefficients
with statistical reliability problems (i.e., vitamins, minerals,
vitamin variance, and mineral variances). Finally, for traditional cereals the estimated coefficients appear much more
reliable with the exception of calories ($p \leq .240$).

To determine if the reason for lack of high level of probability for some of the dietary components is due to lack of
variation within the sample data, means and standard deviations
were inspected for dietary components across cereal types.

The standard deviations and the means for dietary characteristics can be observed in Table 11.2. Although not reported in
this table, the coefficients of variation (CV) have been calculated as a basis for analyzing the effect of variation on the
significance levels of estimated coefficients. Such calcula-
tions revealed especially low coefficients of variation for
calories in the high-nutritive cereals (CV = .021) and in the
"kids' stuff" cereals (CV = .031). Although not as low as these
two, some other coefficients of variation were found to be lower

TABLE 11.2. Estimates of parameters and contribution to price per ounce for
    dietary components of breakfast cereals

Dietary component	Coefficient	Probability level	Mean value of dietary component	Standard deviation of dietary component	Contribution to price per ounce ($)
Protein					
All cereals	0.008	0.001	2.291	1.583	0.018
High-nutritive	0.009	0.001	5.508	2.641	0.050
"Kids' stuff"	-0.004	0.007	1.138	0.545	-0.005
Traditional	0.005	0.001	2.426	0.996	0.012
Vitamins					
All cereals	0.010	0.557	0.113	0.109	0.001
High-nutritive	0.135	0.024	0.338	0.258	0.046
"Kids' stuff"	0.066	0.231	0.119	0.015	0.008
Traditional	0.069	0.133	0.082	0.053	0.006
Minerals					
All cereals	0.067	0.047	0.038	0.039	0.003
High-nutritive	0.281	0.236	0.094	0.041	0.026
"Kids' stuff"	0.063	0.787	0.016	0.009	0.001
Traditional	0.184	0.001	0.041	0.040	0.008
Vitamin variance					
All cereals	0.087	0.191	0.016	0.024	0.001
High-nutritive	-0.267	0.265	0.068	0.060	-0.018
"Kids' stuff"	0.256	0.458	0.014	0.002	0.004
Traditional	-0.473	0.181	0.009	0.007	-0.004
Mineral variance					
All cereals	-0.165	0.002	0.008	0.022	-0.001
High-nutritive	-0.469	0.012	0.050	0.057	-0.023
"Kids' stuff"	-0.157	0.886	0.001	0.002	-0.001
Traditional	-0.503	0.001	0.005	0.011	-0.003
Fiber					
All cereals	-0.006	0.009	0.425	0.492	-0.003
High-nutritive	-0.068	0.208	0.228	0.113	-0.016
"Kids' stuff"	-0.018	0.183	0.148	0.079	-0.003
Traditional	-0.008	0.004	0.583	0.564	-0.005
Calories					
All cereals	-0.014	0.053	1.050	0.102	-0.015
High-nutritive	0.156	0.545	1.066	0.022	0.166
"Kids' stuff"	-0.071	0.021	1.113	0.034	-0.079
Traditional	0.010	0.240	1.018	0.115	-0.010

Source: Morgan, et al. (1979), with permission.

than would have been the case had the data been experimentally
generated. For example, for "kids' stuff" cereals the coefficient of variation was .280 for vitamin C and .174 for vitamin
variance; for calories for traditional cereals and all cereals
combined the coefficient of variation was .113 and .098 respectively. It should be noted that in all groups of cereals,
calories were found to be low in variation when compared to
other dietary components.

Fairly reasonable coefficients have been estimated for the
variables protein, minerals, vitamins, and fiber; however, the
model appears to have limitations for estimation of coefficients
for calories, largely due to the lack of variation in this variable. Even though the developed hedonic index for breakfast
cereals is not yet perfectly formulated (nor is it in related
exploratory research based on survey data), attention will now
be turned to evaluating the contribution of the dietary compon-

ents to the value of one ounce of cereal. This contribution was
determined by multiplying the estimated coefficient for a given
dietary component by the mean value for the same variable. To
determine the total contribution of vitamins and minerals to the
price of breakfast cereals, the resulting contributions of
vitamins (minerals) and vitamin variance (mineral variance) were
combined. For example, for high-nutritive cereals the contribution of vitamin variance (-0.018) was combined with the contribution of vitamins (0.046) to obtain the total contribution of
vitamins (0.028). The contributions to the value of one ounce
of cereal for protein, total vitamin, total mineral, fiber, and
calories were summed to obtain the implicit price of dietary
components in one ounce of cereal. The following implicit
prices were obtained: for all cereals combined the implicit
price of the dietary components in one ounce of cereal was
$0.004; for high-nutritive cereals, $0.233; for "kids' stuff"
cereals, negative $0.073; and for traditional cereals, $0.023.
The implicit price for the dietary components of traditional
cereals seems most plausible, followed by that for all cereals
combined.
     The apparently incongruous results for the classes of high-
nutritive and "kids' stuff" cereals can be attributed, at least
partly, to the coefficients estimated for calories. In the case
of high-nutritive cereals, the implicit price for calories was
found to be $0.166 per ounce of cereal. As previously reported,
this estimate is unreliable because the coefficient of variation
indicated the great lack of sample variation for calories. Without sufficient sample variation, this variable becomes in effect
another constant term. This condition seems to have been the
case for the high nutritive cereals. Comparing the estimated
coefficients for calories and the constant term in high-nutritive cereals indicates that they have offsetting effects. The
constant term (intercept) for high-nutritive cereals was -0.162
which in absolute value is approximately equal to the coefficient for calories multiplied by the average value for calories
in the high-nutritive cereals.
     The problem of lack of variation is also found for calories
in the "kids' stuff" cereals. However, not all of the problem
of the less plausible implicit price for the dietary components
of "kids' stuff" cereals can be attributed to this factor alone.
The level of probability for the coefficients for minerals and
mineral variance is also too high for confidence in the estimates. Additionally, the value for vitamin variance is fairly
high. With this evidence and the fact that "kids' stuff" class
had a considerably smaller R^2 value than the other two classes
of cereals, it can be concluded that the probable reason for the
"kids' stuff" cereal class model being less appropriate than that
for the other two classes of cereals is that dietary components
are of lesser value to "kids' stuff" cereal buyers than to other
cereal buyers. It can be speculated that demand for "kids'
stuff" cereals is more a function of advertising and/or flavor

of cereal than it is of the dietary components.

The effect of the quantitative variable representing length of contract on the price of cereals can virtually be dismissed, since the estimated coefficient is quite small and, in only one case, statistically significantly different from zero. Furthermore, this variable was found not to have made a statistically significant contribution to R^2.

Using the appropriate coefficients obtained for the qualitative characteristics (Table 11.1), conclusions can be drawn as to the combined effect these nondietary characteristics have on the price of one ounce of cereal. A useful way to illustrate how these findings can be interpreted as to their contribution to the price of one ounce of cereal is to consider an example. As a hypothetical situation the cereal to be assessed is defined as follows: Kellogg's Corn Flakes purchased in bulk form through a winning bid contract in the state of Minnesota. For the hypothetical cereal the contribution of all qualitative factors to the price of one ounce would be determined by using the estimated coefficients displayed in Table 11.1. Thus, from the intercept value of -0.024, nothing would be subtracted or added for package size since the coefficient for bulk package appeared in the intercept value; 0.086 and 0.001 would be added to the intercept value since this was a ready-to-eat cereal and was prepared from corn; 0.033, 0.003, 0.002, and 0.022 would be subtracted from the intercept value since these are the negative coefficients obtained for the process of flaking, the winning bid contract, the Kellogg production company, and Minnesota as the state where purchased, respectively. Adding and subtracting all these coefficients to the intercept value results in a value of 0.003. The interpretation is that the combined effect of all nondietary characteristics contributes $0.003 to the price of one ounce of corn flakes.

To further extend this analysis of Kellogg's Corn Flakes, the contribution of the dietary components to the price of one ounce were calculated and found to be $0.025. Thus for the average price per ounce of Kellogg's Corn Flakes, $0.046, this hedonic index model has explained $0.028 of it.

CONCLUSIONS AND RECOMMENDATIONS

This exploratory investigation, while limited by the type of data utilized, has been relatively successful in developing a predictive model for the prices of breakfast cereals. It has established reasonable relationships between price per ounce of cereal and a group of aggregated dietary components. Additionally, this model has revealed reasonable price differentials attributable to other selected breakfast cereal characteristics and market factors related to the purchase of cereals.

Recommendations for continued research in this area include collecting information from the retail market rather than the

institutional market, sampling to permit results that can be
more readily generalized, and extending the model to include
additional characteristics of breakfast cereals. The purpose of
such projects could be twofold. First, comparisons could be
made between the purchasing behavior of institutional buyers and
consumers of retail products. Second, by extending the model to
include some additional characteristics and market factors of
breakfast cereals, more extensive information could be obtained
about buyer behavior in the specified market settings. Variables describing advertising expenditures and flavor and
texture of cereals might be included. The current exploratory
results suggest that this step in applying the commodity
characteristics-based theory of consumer demand is promising.

Finally, similar models might be developed for other food
products and compared to the present results. Such comparisons
would be helpful in indicating whether consumers behave consistently in the implicit pricing of nutrients across products and
also in studying the influence of relative prices of dietary
components in various foods on consumption patterns.

NOTES

1. The 19 components based on ingredients in one ounce of
cereal are protein, vitamin A, vitamin C, thiamin, riboflavin,
niacin, iron, vitamin D, vitamin B_6, folacin, phosphorus, magnesium, zinc, calcium, copper, vitamin B_{12}, vitamin E, calories, and
crude fiber. Five other product characteristics include package
size (individual serving or bulk), preparation required (ready-
to-eat or needs-to-be-cooked), class of cereal (high-nutritive,
"kids' stuff," or traditional), type of processing (puffed,
flaked, shredded, granulated, oven-toasted, or milled), and
principal grain (corn, wheat, oats, rice, barley, or mixed).
Four market factors include type of arrangement through which
price was quoted (winning bid, losing bid, or negotiated),
length of contract, manufacturer (General Foods, General Mills,
Kellogg, Quaker Oats, Ralston Purina, Standard Milling, or Van
Brode), and state where purchase was made (Arkansas, Florida,
Hawaii, Maine, Massachusetts, Minnesota, Missouri, Nebraska,
North Dakota, Oklahoma, Virginia, or Washington).

2. Linear, semilogarithmic, and polynomial functional forms
were estimated. The linear form was chosen on the basis of
explanatory power and consistency of signs for the estimated
structural coefficients.

3. States generally participated if they had centralized
purchasing for the institutions using cereal products and their
records were kept in readily accessible form or the cost of
providing the requested information was nominal. The fact that
the state effect proved significant in some cases would suggest
caution in generalizing the results. On the other hand, the

states complying with the request for data appeared to have a reasonable geographic distribution.

4. An alternative approach would have been to use the raw data and weight the observations for the desired representative effect. This procedure, while statistically more appropriate, was cumbersome. In view of the exploratory nature of the undertaking, the simpler alternative was selected.

5. The number of observations for the time periods ranged from 168 for 1973-74 to 408 for 1975-76. The different classes of cereals were approximately equally represented--231 observations on the seven specific high-nutritive cereals, 732 observations on the 26 "kids' stuff" cereals, and 739 observations on the 43 traditional cereals. These data provided a reasonable representation for assessing the relationship of dietary characteristics to price. For the other nondietary product characteristics and market factors, the sample was less representative: 1,104 observations for negotiated contracts as compared to 598 bid contracts and 684 observations for bulk packaging compared to 1,018 for individual servings. As the focus of the analysis was on the dietary characteristics, these somewhat unequal distributions across the other nondietary product characteristics and market factors were considered acceptable.

6. High-nutritive cereals were defined as the cereals that are highly supplemented with nutrients. "Kids' stuff" cereals are cereals that have been marketed in such a way as to appeal to children; they are usually sugar coated. Traditional cereals are all cereals not included in the two other categories.

REFERENCES

Adelman, S., and Z. Griliches. 1961. On an index of quality change. J. Am. Stat. Assoc. 56:535-38.
Baily, M. J., R. F. Muth, and H. D. Nourse. 1963. A regression method for real estate price index construction. J. Am. Stat. Assoc. 58:933-42.
Brim, O. J., D. C. Glass, D. E. Lavin, and N. Goodman. 1962. Personality and decision process. Stanford, CA: Stanford Univ. Press.
Brown, S. L. 1964. Price variation in new houses, 1959-1961. USDC, Bur. Census Staff Work. Pap. Econ. and Stat. No. 6. Washington, DC: U.S. Government Printing Office.
Cagan, P. 1965. Measuring quality change and the purchasing power of money: An exploratory study of automobiles. Nat. Banking Rev. 3:215-36.
Cohen, J. B., M. Fishbein, and O. T. Ahtola. 1972. The nature and uses of expectancy-value models in consumer attitude research. J. Mark. Res. 9:456-60.
Dean, C. R., and H. J. DePodwin. 1961. Product variation and price indexes: A case study of electrical apparatus. Proc. Bus. Econ. Stat. Sect., Am. Stat. Assoc. Washington, DC.

Dhyrmes, P. J. 1967. On the measurement of price and quality changes in some consumer capital goods. Am. Econ. Rev. 57:501-11.

Federal Trade Commission. 1976. A survey of consumer responses to nutrition claims. Response analysis. Princeton, NJ.

Fetting, L. P. 1963. Adjusting farm tractor prices for quality changes, 1950-1962. J. Farm Econ. 45:599-611.

Fisher, F. M., Z. Griliches, and C. Kaysen. 1967. The costs of automobile changes since 1949. J. Pol. Econ. 70:433-51.

Gavett, T. W. 1967. Quality and piece price index. Mon. Labor Rev. 90:16-20.

Gillingham, R., and D. C. Lund. 1970. A hedonic approach to rent determination. Proc. Bus. and Econ. Stat. Sec., Am. Stat. Assoc., Washington, DC.

Griliches, Z. 1968. Hedonic price indexes for automobiles: An econometric analysis of quality change. In Readings in economic statistics and econometrics, ed. A. Zellner. Boston: Little, Brown.

Kravis, I. B., and R. E. Lipsey. 1969. International price comparisons. Int. Econ. Rev. 10:233-46.

Ladd, G. W., and M. B. Martin. 1976. Prices and demands for input characteristics. Am. J. Agric. Econ. 58:21-30.

Ladd, G. W., and V. Suvannunt. 1976. A model of consumer goods characteristics. Am. J. Agric. Econ. 58:504-10.

Lancaster, K. 1966. A new approach to consumer theory. J. Polit. Econ. 74:132-57.

Mazis, M. B., O. T. Ahtola, and E. R. Klippel. 1975. A comparison of four multi-attribute models in the production of consumer attitudes. J. Consum. Res. 2:38-52.

Morgan, K. J., E. J. Metzen, and S. R. Johnson. 1979. An hedonic index for breakfast cereals. J. Consum. Res. 6:67-75.

Musgrave, J. C. 1969. The measurement of price changes in construction. J. Am. Stat. Assoc. 64:771-86.

Schwartz, N. E. 1975. Nutritional knowledge, attitudes, and practices of high school graduates. J. Am. Diet. Assoc. 66:28-31.

Triplett, J. E. 1963. Automobiles and hedonic quality measurement. J. Polit. Econ. 77:408-17.

U. S. Department of Agriculture. 1975. Homemakers' food and nutrition knowledge, practices, and opinions. Home Econ. Res. Rep. No. 39. Washington, DC.

Waugh, F. V. 1927. Quality as a determinant of vegetable prices. New York: Columbia Univ. Press.

CHAPTER 12

Food Consumption and Nutrient Intake Patterns of School-Age Children

FOOD HABITS in the U.S. have undergone profound changes in recent years as the result of increased urbanization, greater mobility, altered life styles, and greater affluence (Parrish 1971). Some researchers believe the changes in food habits have led to diet deterioration (Hodges and Krehl 1965; Callahan 1971; Parrish 1971) with a particularly strong impact on the diets of school-age children.

In the first part of this chapter Morgan reports on research relating to three specific issues of children's food habits: (1) the impact of ready-to-eat breakfast cereal on nutrient intake, (2) the sources of sugar, and (3) the impact of salted snack foods.

The effects of two food delivery programs, the school lunch and the food stamp program, on the nutrient intake of children have in recent years been of considerable interest to both policymakers and the general public. Price reports on the effects of these delivery programs on the nutrient intake and food consumption patterns of 8 to 12-year-old school children in the state of Washington. The effects of other socioeconomic variables are also included in the analysis. Of particular interest are the effects of coming to school without breakfast, income and assets, ethnic group, and household size.

ISSUES CHALLENGING THE FOOD INDUSTRY

Recently, various segments of the food industry have been challenged about the distribution of their food products to children.[1] For example, the Federal Trade Commission has challenged the ready-to-eat breakfast cereal industries on the impact of cereals on children's diets.

Over the past decade, the regulation of the sale of foods of minimal nutritional value on school premises has undergone much

scrutiny. The most recent ruling on the competitive foods issue
has stated that foods not to be sold at school are those providing less than 5 percent of the Recommended Daily Allowance
(RDA) per 100 calories and less than 5 percent of the RDA for
each of eight specified nutrients per serving (Eisenman and
Longen 1980).[2] Though thus far the USDA has determined only
twelve foods to be banned from school, industry awaits an amend-
ment to the competitive food rule. Changes in the rule can, of
course, have tremendous impact on various food industries such
as snack food producers, candy producers, soft drink producers,
and so forth.
     Economists are well versed in assessing the impact of such
federal regulations on private industry. However, the first
question to address is whether such regulations are truly needed. It is important to examine data which illustrate the impact
of consumption of various foods on diet quality before developing analyses which project impact on sales due to the curtail-
ment of these consumption patterns.

Methods
     Market Facts of Chicago, Illinois, gathered the data used in
this investigation. Seven-day food diaries were mailed during
the third week of September 1977 to 2,000 of the firm's Consumer
Market Panel II families which were balanced by geographic area,
population density, degree of urbanization, income, and age of
panel member. Diaries of 1,434 families were returned and considered to be in usable form. From these diaries, observations
of food intake of children ages 5 to 12 years (657 children from
404 families) were coded for computerized analyses.
     Analysis of the sample characteristics indicated it was
representative of a cross-sectional sample of middle- to upper-
middle class, two-parent families composed of four to six members
with parents who have attained an educational level somewhat
higher than the general U.S. population (Morgan et al. 1980).
     Utilizing the Michigan State University nutrient data bank,
the food consumption data for seven recorded days per child was
analyzed for average daily nutrient intake. Results indicated
the majority of the sample children consumed adequate amounts of
most of the 25 nutrients for which evaluation was done; however,
frequent low intakes were found for vitamin B_{12}, calcium, magnesium, and zinc (Morgan et al. 1980).

Impact of Ready-to-Eat Breakfast Cereal
     The sample of 657 children were partitioned into five groups.
The first three groups were composed of children who ate three
or more breakfasts containing specified items during the survey
week. More specifically, for the first three groups, the children were classified as eaters of presweetened ready-to-eat
cereal (PSRTE, n = 177), eaters of nonsweetened ready-to-eat
cereal (NSRTE, n = 150), and eaters of any ready-to-eat cereal

(any RTE, n = 349). Thus, a breakfast counted as a ready-to-eat cereal breakfast might also contain bacon and eggs or it might contain only juice in addition to the cereal. Children could be classified as being in either one or all three of these groups. The fourth and fifth groups, eaters of nonready-to-eat cereal, were composed of children consuming ready-to-eat cereal less than three times at breakfast (< 3 RTE, n = 308) and children who consumed no ready-to-eat cereal at breakfast (no RTE, n = 92). Thus, children could be classified in both < 3 RTE and no RTE groups if they consumed no ready-to-eat cereal at breakfast. Consumption of ready-to-eat cereal at other times of day was disregarded for classification of children. An average nutrient intake and percentage of National Research Council-Recommended Daily Allowance (NRC-RDA) for each group was calculated for the breakfast meal and for the total day. Additionally, each group was assessed for the number of times a breakfast was skipped during the seven-day period. The t-tests were employed to test for statistically significant differences in nutrient intakes between groups of children (i.e., PSRTE vs. NSRTE, any RTE vs. no RTE, and any RTE vs. < 3 RTE).

Average total protein intake at breakfast was similar for all groups with a significant variation ($p \leq .01$) found only between eaters of PSRTE and NSRTE cereal. Highly significant differences in fat consumption were found when no RTE and < 3 RTE cereal groups were compared with any RTE cereal group. Children classified as eaters of breakfast cereals (any RTE) were found to have consumed significantly greater ($p \leq .05$) amounts of carbohydrate and total sugar at the breakfast meal than did those children who never ate ready-to-eat cereal at breakfast. Cholesterol intake at breakfast was significantly greater for children not consuming any ready-to-eat cereal. The group of children considered to be eaters of cereal had significantly higher intakes for all vitamins and minerals at breakfast than did individuals who consumed no cereal with the exceptions of sodium and zinc. Explanation for these higher intakes by eaters of cereals lies in the fact that nearly all cereals are fortified with many of these nutrients. The higher intakes of vitamin D and calcium by the eaters of cereal is evidenced by the fact that in 97 percent of the times cereal was consumed for breakfast, it was consumed with milk.

When children who ate PSRTE cereal three or more times for breakfast during the one week's observation were compared with those who consumed a similar pattern of NSRTE, the PSRTE cereal group was found to have consumed significantly less protein, cholesterol, crude fiber, folacin, phosphorus, sodium, potassium, and magnesium.

Cooksey and Ojemann (1963) and Heseba and Brown (1968) reported that availability of ready-to-eat foods was an important factor in deciding whether or not to eat breakfast. Their conclusions are supported by this research. Analysis showed that for the total sample (i.e., 4,599 possible breakfast meals), 2.6

percent (118) of possible breakfasts were skipped. When this assessment was made for each of the five groups, it was found that for children considered to be eaters of PSRTE cereals, 1.0 percent of possible breakfast meals were skipped; for NSRTE, 1.0 percent; for any RTE, 1.2 percent; for < 3 RTE, 4.2 percent; and for no RTE, 5.6 percent.

Using the same classification system for the children as used when studying breakfast consumption patterns, average total daily intakes were assessed for each of the five classifications as well as for total sample. For most classifications and for total sample, the RDA, for nutrients which have established RDA, were nearly met and in many cases exceeded.

Children who consumed ready-to-eat cereal three or more times at breakfast during the observed week obtained, on the average, a significantly greater average daily intake of certain nutrients than children who did not consume ready-to-eat cereal. Those nutrients included thiamin, niacin, riboflavin, pyridoxine, vitamin B_{12}, folacin, vitamin D, and iron. On the other hand, the children who did not consume ready-to-eat cereal consumed significantly greater quantities of cholesterol.

When comparisons of average daily intakes for eaters of PSRTE cereal and eaters of NSRTE cereal were made, results indicated eaters of NSRTE cereal obtained significantly greater amounts of protein, cholesterol, crude fiber, vitamin B_{12}, folacin, total vitamin A, calcium, phosphorus, potassium, and magnesium.

In summary, this investigation has shown that, on the average, children ages 5 to 12 years consumed adequate breakfasts, that is, breakfasts contributed at least one-fourth of NRC-RDA for all nutrients. Further, these findings indicate that, in general, ready-to-eat cereals made a contribution to the nutrient consumption of children ages 5 to 12 years. In addition, average daily intakes of total sugar indicated that children who ate ready-to-eat cereal at breakfast consumed, on the average, only slightly more total sugar than eaters of nonready-to-eat cereal and did not consume significantly more total sugar than children who did not consume cereal.

Sources of Sugars in Children's Diets

A large number of studies (Critchley et al. 1967; Newbrun 1967; Carlson and Sundstrom 1968; Fry and Grenby 1972; Bibby 1975, 1977; Edgar et al. 1975)) have addressed possible relationships between the ingestion of simple carbohydrate and dental caries. However, no comprehensive study has been executed to assess the sources from which the majority of children obtain the sugars they eat.

The previously described sample of 657 children ages 5 to 12 years was used to investigate the food sources of total sugar[3] consumption by children. Both average consumptions and distributions of consumptions were assessed for thirteen food groups.

Results from analyses of average daily consumption of total sugar are provided in Table 12.1. A further description of sources of total sugar consumed by children was obtained through frequency distributions. Such analyses showed the large majority of children obtained rather small amounts of total sugar, on an average daily basis, from the ingestion of sweetened dairy products and frozen confections; donuts and sweet rolls; crackers, breads, and waffles; breakfast cereals; candy; jellies, syrups and table sugar; and all other foods. These frequencies showed that a few children obtained sizable quantities of sugars from particular food groups even though the majority of children did not. For example, a few children consumed jellies, syrups, and table sugar at such a level as to obtain nearly 65 gm from it during the average day even though the majority of the sample children received about 7 gm of sugars per average day from the consumption of jellies, syrups, and table sugar. Similar findings are seen for the food groups donuts and sweet rolls, candy, and all other foods.

TABLE 12.1. Analysis of average daily consumption of total sugar by children 5 to 12 years old

Food group	Mean and S.D.[a] (gram)	Percentage of total sugar
Breakfast cereals	4.2 ± 4.6^{fg}	3.3
Cakes, cookies, pies, other desserts	15.3 ± 13.4^{c}	11.2
Candy	3.7 ± 6.2^{g}	2.6
Crackers, breads, waffles	5.2 ± 4.0^{f}	4.1
Dessert sauces	4.5 ± 13.6^{fg}	2.5
Donuts, sweet rolls	2.2 ± 4.2^{h}	1.6
Fruit	17.1 ± 30.1^{b}	11.5
Fruit juices	12.0 ± 14.6^{d}	8.8
Jellies, syrups, table sugar	12.8 ± 11.0^{d}	9.8
Milk	25.9 ± 14.0^{a}	20.4
Sweetened beverages	17.9 ± 16.4^{b}	13.8
Sweet dairy products, frozen confections	7.4 ± 8.4^{e}	5.5
Other food items	6.1 ± 3.4^{f}	4.9
All foods (total)	134.3 ± 48.1	100.0

Note: Total sugar is composed of all simple sugars including glucose, sucrose, lactose, maltose, and other reducing saccharides.

[a]The letters are used to indicate the results of Duncan's multiple range test for group means (Duncan 1957). The same letter identifies where the average daily total sugar consumption obtained from the food groups are not statistically different from each other at the .05 significance level.

Very different findings were obtained for the food groups milk; cakes, cookies, pies, and other desserts; and fruit. These food groups contributed to a more normal distribution pattern of total sugar intake. Although tails were obtained at the upper end of the distributions, thus indicating children consumed large quantities of some of these items, most children exhibited consumption patterns of near normal distributions of these foods. Somewhat different distributions were obtained for the remaining three food groups—sweetened beverages, fruit juices, and dessert sauces. For these food groups, many children obtained approximately 4 gm of sugars from eating the various items; however, quite a few children also obtained large quantities of sugars (80 to 100 gm) from consumption of them.

Although this study, which is reported in depth by Morgan and Zabik (1981), has revealed total sugar consumption patterns for children and has indicated probable sources of this total sugar intake, no attempt has been made to relate this consumption to the occurrence of dental caries for several reasons: (1) no distinction has been made between time of consumption (i.e., meal versus between meal consumption), (2) no recognition has been made of the retention of the varying sugars on the surfaces of the teeth, and (3) no distinction has been made between the relative cariogenicity of liquid versus solid sugars. All of these factors have been found to impact on caries activity (Gustafsson et al. 1954; Glass and Fleisch 1974; Bibby 1975, 1977; Edgar et al. 1975). Additional studies in this area are needed to assess whether or not a relationship exists between diet and the occurrence of dental caries.

Selected Salted Snack Food Consumption

Some nutritionists have expressed concern about abusive usage of salted snack foods (SSFs) such as potato chips by school-age children. However, other nutritionists have contended that specific foods in the diet, including salted snack foods, are neither good nor bad. Thus, the objective of the research reported herein was (1) to evaluate these two viewpoints by examining the food consumption patterns of 5- to 12-year-olds as they relate to the intake patterns of selected salted snack foods and (2) to determine whether consumption of such foods leads to dietary inadequacies.

The previously described sample of children ages 5 to 12 years was partitioned to reflect groups of children who consumed various amounts of selected salted snack foods. The sample was divided into five groups. Group one (n = 16) consisted of all children who obtained greater than 10 percent of their average daily caloric intake from the consumption of SSFs. Those children who obtained less than 10 but greater than 5 percent of their average daily caloric intake from the SSFs were placed in another group (n = 65). This classification procedure was followed in dividing remaining children into categories which reflected their consumption levels of SSFs as greater than 3 percent but less than 5 percent of total caloric intake (n = 104), greater than zero but less than 3 percent of total caloric intake (n = 361), or as zero—no SSFs consumed during the one week surveyed (n = 111). For each of the five classified groups of children, average daily intake of 24 dietary components were calculated as well as standard deviations and average percentage

NRC–RDA.   Each   set   of   mean   intakes   were   analyzed   for
statistically significant differences.

Approximately 83 percent of all children in the sample consumed SSFs. Those children who consumed SSFs consumed an
average of 3.3 ounces per week. Average consumption per week
for all 5- to 12-year-olds in sample was 2.5 ounces. Analysis
revealed an increase in the amount of SSFs consumed as the age
of the children increased: the average intake for 5- and 6-year-
olds was 2 ounces per week and for 11- and 12-year-olds, 3.1
ounces per week.

Mean intakes and statistical comparisons for the various
groups of children are reported in Table 12.2.[4] These data,
along with an in-depth report by Morgan et al. (1980), demonstrate that the consumption of varying proportions of energy
from SSFs had little impact on the average daily nutrient intake
of children 5-12 years old. In general, the intake of all nutrients was adequate with the possible exceptions of magnesium and
zinc, the intakes of which were marginal. However, it should be
noted that the proportion of calories derived from SSFs did not
have an impact on the intake of these two nutrients. Thus, the
generally held notion that the consumption of SSFs has a detri-
mental effect on the nutrient intake of children is simply not
borne out by the results of this study. The data, in fact, show
children consuming SSFs had an average nutrient intake quite
comparable to that of children not consuming these foods.

IMPACT OF SOCIOECONOMIC FACTORS, FOOD
DELIVERY PROGRAMS, AND OTHER VARIABLES

There has been limited research on the influence of socioeconomic factors and food delivery programs on the nutrient
intake of children.[5] Adrian and Daniel (1976), using 1965 USDA
data, found income to be a significant positive factor affecting
families' consumption of all nutrients except carbohydrate.
They also found that families with more educated women consume
less carbohydrate and fat and more vitamin C. This finding
coincides with that of Haley et al. (1977). Eppright et al.
(1969) reported that the educational background of parents
strongly influenced food habits of grade school children.
Fusillo and Beloian (1977) also found low nutrition knowledge
and poorer eating habits for individuals with less education,
lower income, and less prestigious occupations. Burke (1961)
found that other social and economic factors such as age, race,
stage in the family life cycle, family size, meal adjustment,
degree of urbanization, and employment status of the homemaker
impact nutrient intake.

Au Coin et al. (1972) reported that sex and educational
background of parents exhibit a strong influence on the food
intake of students; however, family size per se was not an influ-
ential factor. Coons (1952) observed that underconsuming groups

TABLE 12.2. Average daily nutrient intake of various groups of children classified according to the percentage of caloric intake from the consumption of selected salted snack foods

| | | | | Percentage of calories from selected snack foods | | | | | | | | | | | |
| Nutrient | 10 (n = 16) | | | 5.0 to 9.9 (n = 65) | | | 3.0 to 4.9 (n = 104) | | | 0.1 to 2.9 (n = 361) | | | 0 (n = 111) | | |
	Ave.	S.D.	Ave. % RDA	Ave.	S.D.	Ave. % RDA	Ave.	S.D.	Ave. % RDA	Ave.	S.D.	Ave. % RDA	Ave.	S.D.	Ave. % RDA
Calories	2220[ab]	529	92[ab]	2136[ab]	566	87[b]	2121[ab]	476	88[b]	2239[a]	581	100[a]	2051[b]	560	92[ab]
Total protein, g	70[b]	19	176[b]	72[b]	19	188[b]	74[b]	17	194[b]	80[a]	21	220[a]	76[ab]	21	211[ab]
Total fat, g	102[a]	27	--	92[a]	31	--	92[a]	23	--	96[a]	29	--	87[a]	24	--
Total carbohydrate, g	264[ab]	67	--	264[ab]	68	--	259[ab]	61	--	274[a]	71	--	248[b]	80	--
Total sugar, g	123[b]	41	--	134[ab]	42	--	132[b]	43	--	145[a]	48	--	130[b]	54	--
Cholesterol, mg	282[ab]	83	--	271[b]	106	--	316[a]	116	--	328[a]	132	--	325[a]	123	--
Crude fiber, g	3.2[ab]	1.5	--	3.1[ab]	0.9	--	2.9[b]	1	--	3.3[a]	1.3	--	2.7[b]	1.1	--
Ascorbic acid, mg	124[a]	107	302[a]	102[a]	52	248[a]	98[a]	48	238[a]	113[a]	64	274[a]	99[a]	65	241[a]
Thiamin, mg	1.23[b]	0.39	113[a]	1.29[ab]	0.33	116[a]	1.24[b]	0.26	113[a]	1.36[a]	0.38	117[a]	1.27[b]	0.40	116[a]
Niacin, mg	17.3[a]	4.5	119[a]	17.2[a]	4.3	117[a]	16.7[a]	3.7	113[a]	17.7[a]	5.1	115[a]	16.8[a]	5.1	114[a]
Riboflavin, mg	1.71[b]	0.60	129[c]	1.94[b]	0.57	144[bc]	2[b]	0.50	148[b]	2.22[a]	0.69	157[a]	2.01[b]	0.57	150[ab]
Pyridoxine, µg	1186[b]	326	93[b]	1302[ab]	377	108[ab]	1263[b]	326	106[ab]	1381[a]	456	115[a]	1266[b]	432	108[ab]
Vitamin B_{12}, µg	3.84[b]	2.72	169[b]	3.84[b]	2.02	185[b]	4.21[b]	2.14	206[b]	5.23[a]	4.56	255[a]	4.33[b]	2.39	218[ab]
Total vitamin A, IU	5006[b]	3003	147[b]	5065[b]	1874	150[b]	5316[b]	2938	158[b]	6359[a]	3843	190[a]	5792[ab]	3091	176[ab]
Iron, mg	12.6[ab]	3.9	103[b]	12.7[ab]	3.5	111[ab]	12.5[b]	2.7	109[b]	13.5[a]	3.8	119[a]	12.5[b]	4.0	111[ab]
Calcium, mg	892[b]	404	97[c]	992[b]	345	112[c]	1035[b]	306	117[bc]	1147[a]	381	131[a]	1055[b]	324	122[ab]
Phosphorus, mg	1228[b]	372	132[b]	1276[b]	364	144[b]	1327[b]	319	150[b]	1452[a]	412	165[a]	1349[b]	366	156[ab]
Sodium, mg[d]	3098[a]	786	--	2907[a]	769	--	2830[a]	733	--	2923[a]	819	--	2636[a]	764	--
Potassium, mg	2546[b]	860	--	2534[b]	672	--	2621[b]	636	--	2856[a]	862	--	2580[b]	825	--
Magnesium, mg	235[b]	76	90[b]	235[b]	60	92[b]	247[b]	81	97[b]	286[a]	123	113[a]	239[b]	72	96[b]
Copper, µg	1482[ab]	540	--	1329[b]	578	--	1423[ab]	512	--	1548[b]	637	--	1354[b]	464	--
Zinc, mg	10.7[ab]	5.1	90[b]	10.4[ab]	4.7	95[ab]	9.9[b]	2.9	89[b]	11.4[a]	4.5	103[a]	10.2[b]	3.7	93[b]

a,b,c Means in a row sharing a common superscript (a, b, c) are not significantly different (p ≤ .05) (Duncan 1957).
d Sodium consumption is underestimated because added table salt not included in calculation.

were most likely to be found among low-income families of large
size with homemakers of less than high school education. Additionally, small families with high incomes and college
educated homemakers often had diets that did not meet nutritional allowances.

In this section the relationships between nutrient intake
and characteristics of children and their families are estimated
with comprehensive models which also include food delivery programs. Relationships between consumption of specific foods and
these characteristics were also made. More detailed results are
reported in Price et al. (1978) and in Price and Price (1982).

The Sample

A survey of 8- to 12-year-old Washington State school children was conducted during 1972 and 1973. The sample plan included three stages: (1) selecting school districts, (2) selecting
schools within districts, and (3) selecting students within
schools. Districts were selected by stratifying by geographic
area. Within an area, districts were selected by probability
sampling with higher probabilities given to districts with high
proportions of black and chicano children. Schools within districts were selected by a similar probability sampling. Within
schools, the goal was to sample equal numbers of children in
each of three ethnic groups, two poverty levels (above and below
125 percent of poverty), and those participating and not participating in the school lunch program. Children within each of
the 12 cells were selected at random. All cells were not fill-
ed, resulting in unequal cell numbers.

The analysis of food consumption included one district which
did not participate in the school lunch program but the analysis
of the nutrient intake data did not. This district included all
white children, so the ethnic proportions differ between the two
analyses.

Children's consumption data were obtained with three 24-hour
recalls. These were spaced over time so that one recall included a weekend while the other two included week days. The content of the meal eaten at school was recorded by trained observers and used as a reference point along with food models for
the recall of other meals and snacks.

Results

Regression models were used to relate various socioeconomic
variables to the intakes of 10 nutrients and to the consumption
of 20 food groups. The 10 nutrients were selected on the basis
of their importance to children's nutrition. The 20 food groups
were those consumed by the largest number of children.

School Lunch Participation. Lunch participation included three
variables: full participants (4-5 times per week), partial participants (2-3 times per week), nonparticipants (0-1 time per
week). For the partial participants, consumption of some items

were significantly below those of either nonparticipant or full participants. An interaction variable showed this to be the case for male participants but not for female.

The sample for the nutrient intake analysis included only schools that had a lunch program in operation. This leads to the question of selectivity bias. For example, if the nonparticipants consume less than the participants for some reason unrelated to the measured variables, the effect of the lunch program would be biased upwards.

One possible way to avoid such bias is to also include children in districts which do not participate in the lunch program. Such districts should be selected to be comparable to the participating districts. One such district was included in the analysis of food consumption. Children from the single nonparticipating district had a higher consumption of some items than either the full or the nonparticipants. This points up the need for extreme caution in selecting districts for comparison. Significant consumption differences are possible within relatively small geographic areas.

One indicator of selectivity bias is the difference between participating and nonparticipating students with respect to consumption of food items not served at lunch. Differences in food consumption between full participants and nonparticipants occurred mainly among items associated with the school lunch and the substitute sack lunch. Absence of significant differences among other items indicates that substantial selectivity bias does not exist.

Full participants had higher intakes of 5 of the 10 nutrients than either partial or nonparticipants. These nutrients were protein, calcium, phosphorous, vitamin A, and riboflavin. Nonparticipants had higher intakes of iron than did others. Full participants had a higher consumption of fluid milk, mixed dishes, canned vegetables, fats and oils, and sauces and toppings than nonparticipants. Consumption of fresh fruit, bread and rolls, and cooked cereal was significantly higher among nonparticipants.

Most consumption differences are related to the differences between the school lunch and the sack lunch. However, the reasons for differences in milk consumption and cooked cereal consumption are not obvious. In most schools, students who bring sack lunches have access to milk which is sold by the school at low cost. Milk consumption by children in the nonparticipating district, which had the special milk program, was equal to that of full lunch participants. Therefore, the lower milk consumption of nonparticipants is not due to the physical nonavailability of milk. It may be due to the desire of students who bring sack lunches to get to the playground early and avoid standing in line just to get milk. The higher consumption of cooked cereal by nonparticipants may be due to the mother wishing to compensate for the lack of a "hot" lunch. Consumption of cooked cereal by students in the nonparticipating district was

closer to that of nonparticipants than to that of full participants. These differences could also be due to selectivity bias,
but the lack of differences among other food items gives credence
to the above explanation.

Most of the increases in nutrient intake attributable to the
school lunch program stem from the higher milk consumption by
full participants. Simple correlations between milk consumption
and intakes of protein, calcium, phosphorus, and riboflavin
ranged from .61 for protein to .86 for calcium. The higher
intake of vitamin A cannot easily be traced to any particular
food item. Nutrient intakes of students from the nonparticipating district were comparable to those of full lunch participants.
This implies that from a strictly nutrient intake basis children
could be as well fed by participating in the special milk program
and carrying sack lunches as by participating in the school
lunch program. Before advocating such a policy, one would need
additional evidence. This survey included only a single nonparticipating district in a town of about 17,000 persons. These
results cannot be generalized to areas of severe poverty or
large metropolitan areas. In such areas it is possible that
some students would not bring a sack lunch if no school lunch
were provided. In addition, the relatively high Aid to Families
with Dependent Children (AFDC) payments in Washington State also
make generalization to states with lower AFDC payments questionable. AFDC payments in the state of Washington were $63.43 per
recipient during April 1972. This compares with a national
average of $52.16. The range among the 50 states was from
$14.73 to $77.20.

Lack of Breakfast. Eleven percent of the sample came to school
without breakfast. Using weighted data we estimated that, statewide, 7 percent of the white, 12 percent of the black, and 13
percent of the Chicano school children went without breakfast.
Intakes of calcium, phosphorus, thiamin, and riboflavin were
significantly lower for children who did not eat breakfast.
Consumption of fluid milk, ready-to-eat cereal, pastries and
desserts, and a food category including waffles and pancakes
were lower for those children with no breakfast. There was some
evidence that children who did not eat breakfast consumed more
at other meals. Consumption of mixed dishes and fresh fruit was
higher for these children. While these increases did tend to
offset the loss of calories and some other nutrients, it did not
offset the lower intake of the milk related nutrients. Thus, a
school breakfast has the potential for raising intakes of certain
nutrients for those children who do not eat breakfast.

Age and Sex of Child. It is well known that, on the average,
older children eat more than younger children and that males eat
more than females. This study covered only the five-year span
of ages 8 to 12. However, during this span the composition of
the diet varied with age. While consumption of many items in-

creased with age, consumption of ready-to-eat cereals decreased. Consumption of fruit, cooked cereal, and mixed dishes did
not increase with age, while consumption of fluid milk and
vegetables did not increase as much as did the consumption or
other items.

The intake of all nutrients increased with age. The change
in the composition of the diet resulted in the increase of the
energy intake by a greater percentage than 8 of the 9 other
nutrients. Only vitamin C intake increased faster than
energy.[6] An increase of one year in age resulted in a 5.3 percent increase in the intake of energy. In contrast, the intake
of vitamin A increased by 1.4 percent, that of thiamin by 2.4
percent, and that of phosphorous by 2.9 percent.

Consumption patterns likewise varied by sex. Consumption of
many food items was lower for females while consumption of fresh
fruit was significantly higher. Little differences between
sexes existed in the consumption of mixed dishes, canned vegetables, cooked cereal, and pastries and desserts. Females had a
slightly but not significantly higher intake of vitamin C, stem-
ming from the higher consumption of fresh fruit. The intakes of
other nutrients were from 6 to 10 percent lower for females than
for males.

Income and Assets. Income had no significant effect on the
intake of any of the 10 nutrients. Using this data set, the
income elasticity for household food expenditures was estimated
to be only 0.04 (West and Price 1976). One would expect a
higher income elasticity for household food expenditures than
for nutrient intake since expenditures embody services and
convenience aspects that have no effect on nutrient composition.
Among the 20 food items, income significantly affected only 4.
Beef and all meat were positively affected while cooked cereal
and bread and rolls were negatively affected. Income elasticities were 0.08, 0.04, -0.10, and -0.07 respectively.

Assets did affect intakes of some nutrients. Liquid assets
positively affected calcium and phosphorus intake while total
assets positively affected riboflavin and vitamin C. Four of
the 20 food items were significantly affected by liquid assets.
Consumption of major dairy products, fresh fruit, and pastries
and desserts were positively affected, but the consumption of
fats and oils was negatively affected. Thus, perishable products
and the nutrients associated with these products were positively
affected by assets. It is difficult to determine if liquid or
total assets are the causal factor because of a very high degree
of multicollinearity. These findings suggest that households
without assets may be lacking cash reserves at the end of a pay
period. Since food is usually purchased with cash, consumption
of perishables would be reduced. Sufficient stocks of nonperish-
ables apparently exist. Assets may be acting as a proxy for
money management. The household which is security conscious

with respect to financial management will plan its food purchases to even out over the month and have cash reserves for
financial emergencies.

Food Stamp Participation. The food stamp program is a vehicle
to enhance nutritional well-being by increasing the amount of
money available for food expenditures. This data showed the
food stamp program to increase food expenditures (West and Price
1976). Food expenditures were included in early runs of the
nutrient intake model but none were significant at the .10 level.
To test for the possibility that food stamps help remedy the low
cash reserves of some families at the end of the pay period, a
food stamp dummy variable was included in the nutrient intake
models. Again, no coefficients were significant at the .10
level.

This lack of relationship between food expenditures or food
stamps and nutrient intake may be found only in areas where
there is a substantial "floor" under income. High AFDC payments
provide such a floor. Another factor is the effect of more than
one food program. Almost all children whose families were
eligible for food stamps, outside the single nonparticipating
district, were full participants in the school lunch program.
Thus, food stamps may have little or no effect on this group but
may affect persons not participating in the school lunch
program. If AFDC payments and the school lunch program were
eliminated, one would hypothesize a substantial effect on
nutrient intake from food stamps.

Ethnic Group. The regression model showed that among the three
ethnic groups white children have higher intakes of the milk
related nutrients, calcium, phosphorous, and riboflavin than
blacks or chicanos. Chicano children had lower intakes of niacin
and thiamin than other children.

Among the 20 food items, black children had a lower consumption of fluid milk, major dairy products, and bread and rolls
while having a higher consumption of meat. Chicano children had
a lower consumption of beef, mixed dishes, all vegetables,
sweets, jams and jellies, and fats and oils than others. White
children had a higher consumption of fresh vegetables, ready-to-
eat cereal, and sauces and toppings than others. Their consumption of cooked cereal and nonalcoholic beverages was lower than
that of blacks and chicanos.

The ethnic food consumption differences that affect nutrient
intake center around milk and meat. Most other ethnic differences have little effect on nutrient intake. There are substitute foods among ethnic groups with roughly the same nutrient
composition.

The lower intakes of thiamin and niacin may be due to a
combination of food consumption differences. Lower consumption
of beef and ready-to-eat cereals appears to be the major contributing factor.

Geographic Area. Intakes of energy, protein, calcium, phosphorus, and riboflavin were significantly lower for children
whose parents were raised in the southeastern United States
(including Texas). These children were primarily black and
chicano.[7] The analysis of food consumption showed these children
to have significantly lower intakes of fluid milk and fats and
oils. The lower consumption of these two food items coupled
with little accompanying higher consumption of other foods
explains the lower intakes of these children.

Household Size. The regression model showed intakes of 5 of the
10 nutrients to be significantly lower in large households.
They included protein, vitamin A, thiamin, riboflavin, and
niacin. Consumption of fluid milk, fresh fruit, and all fruit
decreased with household size. Decreases in most other food
categories occurred, but these decreases were not significant.
The exceptions to the decreases were breads, biscuits, crackers,
cooked cereal and sweets, jams and jellies. The composition of
the diet therefore changes as household size changes with a
corresponding reduction in the intake of some nutrients. One
apparent source of these reductions is milk consumption. Other
sources also contribute to the lower intakes, but these are not
apparent.

Vitamin Supplements. Children taking a multiple vitamin supplement had higher intakes of all ten nutrients. Since these
supplements usually contain iron and the five vitamins, these
relationships would be expected. However, the increases in
intakes of energy, protein, calcium, and phosphorus are not
directly related to vitamin supplements.
    Analysis of the consumption of the 20 food items shows
higher milk, dairy product, bread and roll, biscuit and cracker,
cooked cereal, and fats and oil consumption among children
taking multiple vitamins. There were no significant increases
in consumption of any of the other food items. The higher
consumption of most food items plus the higher energy intake
indicates these children are generally consuming more food than
others.

Education of Female Head. A variable that may act as a proxy
for the nutritional concern of the mother is education. The
analysis of the 10 nutrients shows consumption of calcium,
vitamin A, and vitamin C to be higher among children from households where the education of the adult female is more than nine
years. Milk consumption was positively related to education
which explains the higher intake of calcium. The only other
food items positively related to education levels were biscuits,
crackers, sweets, jams, and jellies. Consumption of mixed
dishes was negatively related to education level. The higher
intakes of vitamin A and vitamin C thus cannot be readily explained by differences in food consumption.

Occupation of Major Income Earner. The only occupational group
that was significantly related to either the nutrient intake or
food consumption of the child was the armed forces. Approximately 10 percent of the sample consisted of children with at least
one parent employed by the armed forces. Intakes of all 10
nutrients were significantly higher for these children. Consumption of all 20 food items was higher with 9 of these being
significant. Intakes of the five vitamins ranged from 20 to 28
percent higher among children of armed forces personnel while
intakes of the 5 other nutrients ranged from 8 to 14 percent
higher.
        The reasons why these children have higher nutrient intakes
and higher food consumption can only be conjectured. It is an
interesting group for further study. Knowledge of how habits
and nutritional information obtained in the armed forces affect
the nutrition of the children may point to more effective methods
of nutrition education.

Summary
        Concerns about the intake of sufficient amounts of important
nutrients have been the primary focus of this chapter. These
concerns are not the only elements in feeding school children.
Obesity was one of the major nutritional problems in the
Washington State sample. Children are consumers. The taste,
appearance, and variety of foods served and the atmosphere in
which they are consumed should be considered in the context of
children's preferences when recommending policy for the feeding
of children.
        The first part of this chapter investigated the effects of
consumption of ready-to-eat cereal and salted snack foods on
nutrient intake of a sample of 657 children 5 to 12 years old.
Sources of sugar were also investigated. Children eating ready-
to-eat cereal at breakfast consumed only slightly more total
sugar than children not consuming cereal at breakfast, but did
not consume significantly more total sugar during the average
day. Intakes of many vitamins and minerals were higher for
consumers of ready-to-eat cereal than for nonconsumers both at
breakfast and for the entire day. Nonconsumers had higher intakes of cholesterol.
        The three food groups with the largest contributions to
total sugar intake were in order of importance: milk, sweetened
beverages, and fruit. The majority of children obtained rather
small amounts of total sugar from sweetened dairy products and
frozen confections; donuts and sweet rolls; crackers, breads,
and waffles; breakfast cereals; candy; jellies, syrups, and
table sugar. A few children did obtain large amounts of sugar
from some of these food groups. No attempt was made to relate
sugar consumption to dental caries.
        Consumption of salted snack food has been a concern of some
nutritionists. However, this study showed that children consum-

ing these snack foods had an average nutrient intake quite comparable to that of children not consuming these foods.

The second part of this chapter investigated the effects of the school lunch and food stamp programs on the nutrient intake of a sample of about 1,000 Washington State school children 8 to 12 years old. Other socioeconomic variables were also included in the models. Full participants in the lunch program had significantly higher intakes of protein, calcium, phosphorous, riboflavin, and vitamin A than did nonparticipants. With respect to types of foods, full participants had a significantly higher consumption of fluid milk, mixed dishes, canned vegetables, fats and oils, and sauces and toppings, but a lower consumption of fresh fruit, bread and rolls, and cooked cereal than nonparticipants. The higher intakes of 4 of the 5 nutrients by participants is likely due to the higher fluid milk consumption.

About 7 percent of Washington's school children came to school without breakfast. These children had significantly lower intakes of calcium, phosphorous, thiamin, and riboflavin. Correspondingly, these children had a significantly lower consumption of fluid milk and ready-to-eat cereal.

Neither current income or food stamp participation had any effect on the intake of the 10 nutrients. Assets were positively related to intakes of calcium, phosphorous, riboflavin, and vitamin C. Liquid assets were positively related to the consumption of perishable foods such as dairy products and fresh fruit.

Other variables which showed significant negative relationships with the child's intake of one or more nutrients were household size, parents with less than nine years of education, parents raised in the southeastern United States, and black and chicano children.

The first study reported in this chapter indicates the nutrient intake of U.S. children is generally at an acceptable level. Although not discussed in this chapter, the Washington State data confirm this conclusion. Insufficient intake of important nutrients or excessive consumption of salt and sugar are, however, problems of certain subgroups of children. The demographics show black and chicano children from large households with limited assets whose parents have a limited amount of education to be most problematic with respect to intake of certain nutrients. Intake of some nutrients would be still lower if the child missed breakfast. The school lunch has the potential for raising intake of most of the nutrients which are lowered by demographic variables and missing breakfasts. It should be given emphasis in those areas where demographics indicate it is most needed. There is, however, a major qualification in the effect of the school lunch. Four of the 5 nutrients positively affected by the lunch were related to milk consumption. It is not known whether participation in only the special milk program would make children as well off nutritionally as participation in the school lunch program.

NOTES

1. Author of this section is Karen J. Morgan.
2. These 8 nutrients are protein, vitamin A, vitamin C, niacin, riboflavin, thiamin, calcium, and iron.
3. Total sugar is composed of all simple sugars including glucose, sucrose, lactose, maltose, and other reducing saccharides.
4. Average sodium consumption reported here does not include added table salt.
5. Author of this section is David W. Price.
6. An apparent inconsistency appears here since fruit consumption did not increase with age. Another important source of vitamin C, nonalcoholic beverages, did increase with age which partially accounts for the increase in vitamin C.
7. There were substantial numbers of black and chicano children whose parents were raised in the West which prevents this finding from being due to ethnic groups.

REFERENCES

Adrian, J., and R. Daniel. 1976. Impact of socioeconomic factors on consumption of selected food nutrients in the United States. Am. J. Agric. Econ. 58:31-38.

Au Coin, R., M. Haley, J. Rae, and M. Cole. 1972. A comparative study of food habits: Influence of age, sex and selected family characteristics. Can. J. Public Health 63:143-51.

Bibby, B. G. 1975. The cariogenicity of snack foods and confections. J. Am. Dent. Assoc. 90:121-32.

_____. 1977. Food relationship to caries. Proc. workshop on cariogenicity of food, beverages, confections and chewing gum. Am. Dent. Assoc. Health Found, Washington, DC.

Burke, M. C. 1961. Influences of economic and social factors on U.S. food consumption. Minneapolis, MN: Burgess.

Callahan, G. L. 1971. You can't teach a hungry child. School Foodservice J. 25:25-40.

Carlson, J., and B. Sundstrom. 1968. Variations in composition of early dental plaque following ingestion of sucrose and glucose. Odontol. Rev. 19:161-70.

Cooksey, E. B., and R. H. Ojemann. 1963. Why do they skip breakfast? J. Home Econ. 55:43-45.

Coons, C. M. 1952. Family food consumption studies. Public Health Rep. 67:788-96.

Critchley, P., J. M. Wood, C. A. Saxton, and S. A. Leach. 1967. The polymerisation of dietary sugars by dental plaque. Caries Res. 1:112-29.

Duncan, D. B. 1957. Multiple range test for correlated and heteroscedastic means. Biometrics 13:164-76.

Edgar, W. M., B. G. Bibby, S. Mundorff, and J. Rowley. 1975.

Acid production in plaques after eating snacks: Modifying factors in foods. J. Am. Dent. Assoc. 90:418-25.

Eisenman, R., and K. Longen. 1980. Restricting snack foods in schools. Nat. Food Rev., USDA, ESCS, NF-11.

Eppright, E. S., H. M. Fox, B. A. Fryer, G. H. Lamkin, and V. M. Vivian. 1969. Eating behavior of preschool children. J. Nutr. Ed. 1:16-19.

Federal Trade Commission. 1978. Children advertising, proposed trade regulation rule making and public hearing. Fed. Regist., 43:17967.

Fry, A. J., and T. H. Grenby. 1972. The effects of reduced sucrose intake on the formation and composition of dental plaque in a group of men in the Antarctic. Arch. Oral Biol., 17:873-82.

Fusillo, A. E., and A. M. Beloian. 1977. Consumer nutrition knowledge and self-reported food shopping behavior. Am. J. Public Health 67:846-50.

Glass, R. L., and S. Fleisch. 1974. Diet and dental caries: Dental caries incidence and the consumption of ready-to-eat cereals. J. Am. Dent. Assoc. 88:807-13.

Gustafsson, B. E., C. Quensel, L. S. Lanke, C. Lundquist, H. Grahnen, B. E. Bonow, and B. Krasse. 1954. The Vipeholm dental caries study. Acta Odontol. Scand. II:232-364.

Haley, M., D. Au Coin, and J. Rae. 1977. A comparative study of food habits: Influence of age, sex, and selected family characteristics. II. Can. J. Public Health 68:301-06.

Heseba, J., and M. R. Brown. 1968. Breakfast habits of college students in Hawaii. J. Am. Diet. Assoc. 53:334-35.

Hodges, R. E., and W. A. Krehl. 1965. Nutritional status of teenagers in Iowa. Am. J. Clin. Nutr. 17:200-10.

Morgan, K. J., G. A. Leveille, and M. E. Zabik. 1980. The impact of selected salted snack food consumption on school age children's diet. Sch. Foodservice Res. Rev. 5:13-19.

Morgan, K. J., and M. E. Zabik. 1981. Amount and food sources of total sugar intake by children ages 5 to 12 years. Am. J. Clin. Nutr. 34:404-13.

Morgan, K. J., M. E. Zabik, R. Cala, and G. A. Leveille. 1980. Nutrient intake patterns for children ages of 5 to 12 years based on seven-day food diaries. Michigan State Univ., East Lansing, Agric. Exp. Stn., Res. Rep. No. 406.

Newbrun, E. 1967. Sucrose, the arch criminal of dental caries. Odontol. Rev. 18:373-86.

Parrish, J. B. 1971. Implications of changing food habits for nutrition educators. J. Nutr. Ed. 2:140-46.

Price, D. W., D. A. West, G. E. Scheier, and D. Z. Price. 1978. Food delivery programs and other factors affecting nutrient intake of children. Am. J. Agric. Econ. 68:609-18.

Price, D. W., and D. Z. Price. 1982. The effects of school trends participation, socioeconomic and psychological variables on food consumption of school children. Washington State Univ., College of Agric. Res. Cent. XB0912.

U.S. Department of Agriculture. 1974. Federal and state stand-
    ards for the composition of milk products. Agric. Handb.
    No. 51, Washington, DC: U.S. Government Printing Office.
West, D. A., and D. W. Price. 1976. The effects of income,
    assets, food programs, and household size on food
    consumption. Am. J. Agric. Econ. 58:725-30.

CHAPTER 13

Impact of the Food Stamp Program on Food Expenditures and Diet

THE FOOD STAMP PROGRAM (FSP) is designed to provide direct subsidies to low-income households to purchase nutritionally adequate diets. According to the authorizing legislation, a primary goal of the food stamp program is "to raise levels of nutrition among low-income households" (Sexauer 1978).

The FSP was initiated as a pilot program in 1961 with less than 50,000 participants (Table 13.1). Since then, through expansion to all areas of the United States and modification of provisions, participation has dramatically increased. In 1970, major revisions were made in the income eligibility standards and in the value of the stamps. The program emphasis changed from one of supplementing food expenditures to providing sufficient stamps to be able to purchase the USDA Economy Food Plan (Reese et al. 1974). The value of the bonus stamps received by an individual approximately doubled. Participation in the program grew from less than 3 million in 1969 to over 12 million in 1973.

During the 1976–78 period the program had between 16 and 18.5 million participants with a federal subsidy of about $5 billion. During 1979, the federal subsidy exceeded the $6 billion congressional ceiling. Participation in 1980 has risen to nearly 21 million. Participation in the FSP and, thus, program cost have increased substantially over the years. Although the operation and management of the FSP have received much attention, the effectiveness of the FSP in achieving its stated objectives has yet to be determined.

The basic provisions of the program provide a sliding scale of benefits depending primarily on income and household size. Prior to elimination of purchase requirements (EPR) in 1979, recipients purchased a given amount of stamps, the amount received being determined by household size. The amount paid for the stamps varied proportionally with income (adjusted for certain expenses) with the lowest income households paying nothing. The

255

TABLE 13.1. Food stamp program: participation and value, 1961-78

Year	Average monthly participation	Total value of coupons ($1,000)	Federal subsidy ($1,000)
1961	49,640	826	381
1962	142,817	35,202	13,153
1963	225,602	49,876	18,640
1964	366,816	73,485	28,644
1965	424,652	85,472	32,505
1966	864,344	174,232	64,813
1967	1,447,097	296,106	105,550
1968	2,209,964	451,801	173,142
1969	2,878,113	603,351	228,819
1970	4,340,030	1,089,961	549,664
1971	9,367,908	2,713,273	1,522,749
1972	11,109,074	3,308,648	1,797,286
1973	12,165,682	3,883,952	2,131,405
1974	12,861,526	4,727,451	2,718,296
1975	17,064,196	7,265,642	4,385,501
1976	18,548,715	8,700,209	5,326,505
1977[a][b]	17,057,598	8,339,805	5,057,724
1978[a][b]	16,043,361	8,310,918	5,165,209

Source: U.S. Department of Agriculture (1977, 1979).

[a]Preliminary.

[b]Data for fiscal years July 1-June 30 for 1963-76; October 1-September 30 beginning 1977.

difference between the face value of the stamps and the amount paid was defined as the value of the bonus stamps which is in essence the amount of subsidy given recipients. Since 1979, recipients simply receive free a quantity of stamps roughly equal to the value of bonus stamps. For those recipients previously paying a substantial amount for food stamps, the restrictive effect of the program is dramatically reduced. For example, if a recipient household was paying 50 percent of the face value of the stamps, he or she was formerly restricted to using all the face value of the stamps for food. Since EPR, only the value of the bonus stamps is restricted to food use. For those households formerly paying little or nothing for the stamps, EPR would have little effect.

The FSP helps eligible households purchase food through normal marketing channels. It does not restrict the types of foods purchased by recipients (with the exception of alcoholic beverages and certain imported products). The program itself modifies the diets of the recipients only in making it easier for them to financially obtain the types of foods they desire. If individuals are malnourished because of the financial inability to purchase a nutritious diet, the FSP will improve nutritional status of the participants. However, it is possible that recipient households increase food expenditures without affecting the nutrient quality of their diets. It is also possible some eligible households spend the same amount for food after participating in the program than they did before. In addition to evaluating the nutritional impact of the FSP, other issues related to the FSP that are of primary concern to program admin-

istrators and policymakers such as predicting program participation due to changes in economic conditions and population
characteristics and assessing the effects of the FSP on food
expenditures are addressed and investigated in this chapter.

In the first section of this chapter, Senauer identifies
factors that influence FSP participation and analyzes the differences between program participants and nonparticipants. In
the second section, Schrimper examines the effects of the FSP on
food expenditures by using two alternative approaches. One
approach is based on comparison of individual household food
expenditure patterns between participating and nonparticipating
households. The second approach is an aggregative analysis
which examines the effects of the FSP on the total demand for
food. The effects of the FSP on food prices are then estimated
assuming various price elasticities of demand and supply of
food. In the third section, Price reports on the effects of
bonus stamps on the nutritional adequacy and quality of school-
age children's diets in the state of Washington. In the final
section of this chapter, Huang reports on the effects of the FSP
on availability of essential food nutrients to low-income households in the southern region of the United States.

FACTORS EXPLAINING PARTICIPATION IN THE FOOD STAMP PROGRAM

Recent research has revealed much about the factors that
influence FSP participation and the differences between program
participants and nonparticipants.[1] This section draws on four
major research studies: two using county-level data and two
using household data.

Studies Based on County Data

Two studies examined the factors that affect participation
in the FSP with county-level data: one in Minnesota by Sexauer
et al. (1976) and one in North Carolina by Hunter (1980). The
former was a cross-sectional study for the 87 counties in Minnesota in April 1975. The latter was a pooled cross-section and
time-series analysis using monthly county data between July 1974
and December 1977, which yielded 4,200 observations.

The participation rate, defined in this case as the proportion of a county's total population participating in the program,
varied from only 1 percent in Rock County, Minnesota, to 11.5
percent in Cass County in April 1975. In North Carolina, participation rates varied from a low of 1 percent to a high of 21
percent in December 1977. The two studies estimated the relationship between county participation and various economic and
social variables. The Minnesota study used ordinary least
squares regression estimation. The North Carolina study used an
analysis of covariance, in essence, a least squares regression
estimation with dummy variables. Both of these studies drew on
earlier work by Hines (1975) which used national county-level
data to examine factors that affect food stamp participation.

The results of these studies can be applied to predicting future changes in food stamp participation. The findings can also identify counties whose participation levels diverge from the norm. With the Minnesota study, a projected participation rate was developed based on each county's socioeconomic characteristics and the statewide relationship between these factors and participation. With the North Carolina study, a county variable allows each county to have its own intercept. Both approaches indicate those counties where the participation level deviates from the norm that would be expected based on that county's characteristics.

Counties in which participation is lower than the norm for the socioeconomic mix of its population could be scrutinized for factors having an adverse effect on program participation. Counties in which participation is substantially higher than expected might also be studied to further identify factors enhancing program participation.

The economic and social characteristics studied influence both the number of people eligible for food stamp benefits and the proportion of those eligible who actually participate. The economic factors primarily affect the former, and social characteristics, the latter.

Poverty. Not unexpectedly, the number below the poverty level and program participation are strongly related. The Minnesota study found that an increase of 100 persons below the poverty level led to an increase in program participation of 34 persons. The North Carolina study found an even higher effect of 92 additional participants.

Unemployment. For each 100 additional persons who became unemployed, utilization of food stamps increased by 60 persons in Minnesota. In North Carolina the effect was 33 persons.

Income. In Minnesota for a $1,000 increase in average per capita income, food stamp participation fell by 0.6 percent. The North Carolina study found that a 1 percent rise in average real capita income led to a fall in participation of 0.78 percent. The unemployment and income relations indicate the importance of the state of the economy to participation levels.

Welfare Recipients. An increase of 100 persons receiving public assistance in Minnesota added 83 persons to the food stamp program. In North Carolina the increase was 31 participants per 100 welfare recipients. When these studies were done, if a welfare recipient lived in a household in which everyone was receiving public assistance, then he or she was automatically eligible for food stamps.

Minority Groups. Using a simple correlation analysis, the proportion of nonwhites and program participation were positively related. However, this effect was evident primarily because minority groups contain a large proportion of the economically disadvantaged. After accounting for these economic factors in a regression analysis framework, the Minnesota study found minority persons were less likely to utilize food stamps than their basic economic needs would indicate. The North Carolina study found a small positive effect, an increase in participation of 6 persons per 100 nonwhites, after accounting for other factors.

Age. The effect of the proportion of the population over age 65 differed between the two states. The North Carolina study found a strong negative impact for the elderly. This result corresponds with the national pattern of the elderly being reluctant to use food stamps even when in dire need. Surprisingly, the Minnesota results deviated from this pattern, which found the elderly more likely to utilize the program. On the other hand, an increase in the percentage of children in the population had a positive effect on program participation in both states.

Urban Rural Differences. The North Carolina study found that classifying an additional 100 persons as urban residents resulted in about 6 additional participants. The Minnesota study, however, found that the effect was not statistically significant.

The specific numerical effects estimated in these studies are somewhat outdated, since there have been significant changes in the FSP since the completion of these studies. In addition, they related to only two specific states, one in the South and one in the Midwest. However, the direction and rough dimensions of the various impacts are probably still appropriate. Above all though, the methodology is still a viable means of studying program participation effects.

Studies Based on Household Data

Two other studies, which primarily examined the relationship between food stamp participation and food expenditure or consumption, provide further insights into the factors that explain program participation. These two studies were based on household data.

A national study by West (1979) used the 1973–74 diary portion of the BLS Consumer Expenditure Survey. West edited the data to remove observations which were incomplete for key variables and ran consistency checks. After screening, the sample contained 10,514 households of which 587 received food stamps, 5.6 percent of the edited sample. For comparison, 5.78 percent of the U.S. population actually received stamps in 1973. By applying appropriate income and asset tests, nonparticipating households were grouped into FSP eligible and FSP ineligible.

The West et al. (1978) study used data collected from households in Washington State in 1971–73. The sample included only families with 8- to 12-year-olds. The subsample used in this study covered 332 households eligible for food stamps with 196

participants and 136 nonparticipants. The results for this subsample were weighted by the proportion of blacks and chicanos in
the state of Washington to make it representative for the state.

The National Study. In the national study the first comparisons
are between participants and all nonparticipants. As expected,
participating households were larger, 3.4 versus 2.9 persons,
and had lower incomes, $3,468.16 annually versus $11,608.67.
Also, as expected, the proportion of participants was higher in
the South than in the other regions of the country. The South
contained 40 percent of the program's participants and only 30
percent of the nonparticipants. These results are consistent
with the higher incidence of low-income households in that part
of the country. The portion of participants in central city
areas of large SMSAs and rural areas is considerably larger than
for nonparticipants. These locations contain larger proportions
of lower income households. Far fewer participants own their
home or a vehicle.

Participating households were more likely to be black and
headed by a female. Some 59 percent of participating households
were headed by a woman, whereas only 22 percent of nonparticipating households were headed by a woman. Participating households
were headed by less-educated individuals and they worked far
fewer weeks per year on average than nonparticipants.

The comparisons between participants and eligible nonparticipants are even more enlightening. In comparing these two
groups, household size is significantly larger among participants than nonparticipants, indicating the presence of more
children. Total money income was roughly the same for the two
groups. However, the average participating household obtained a
much higher proportion of its income from transfer payments and
a much smaller proportion from earnings than eligible nonparticipants. Income from welfare, alimony, and private pensions
averaged $1,389.40 for participants and only $294.68 for eligible
nonparticipants.

In terms of regional location, the proportions of participants and eligible nonparticipants did not vary significantly.
Eligible nonparticipants were more likely to own their home than
participants, 46.9 percent versus 24.5 percent. They were also
more likely to own an automobile. West found a larger proportion
of participating households headed by women than for eligible
nonparticipants, 59.3 percent versus 40.1 percent. Finally,
these comparisons indicated that blacks were more likely to
utilize the program than other eligibles. Blacks were 39.5
percent of the participants, but only 18.9 percent of the eligible nonparticipants. On the other hand, the elderly were less
likely to participate in the program than younger persons. The
elderly were 24.2 percent of the participants and 34.1 percent
of the eligible nonparticipants.

The Washington State Study. The study in Washington also compar-

ed the socioeconomic characteristics of participating and eligible nonparticipating households. Total monthly income on an
adult equivalent basis was at $115.32 for participants and at
$112.96 for eligible nonparticipants. However, as in the nation-
al study, nonparticipants received most of their income from
earnings, whereas participants received most of theirs from
transfer payments. The mean earnings of the major income earner
was $279.42 in eligible nonparticipating households and only
$78.05 in participating households. On the other hand, the
average participating household received welfare payments of
$192.47 per month and the nonparticipants only $33.69.

Liquid assets, also on an adult equivalent basis, averaged
nearly twice as great for the nonparticipating eligibles, at
$75.40, as for the participants, at $45.50. The value of the
home of eligible nonparticipants was $8,826.00, over twice that
for participants, whose home on average was worth $3,584.00. As
in the national study, this difference was largely due to a much
smaller portion of participants owning their homes. The monthly
housing costs, however, averaged about the same for the two
groups. The value of the vehicle for eligible nonparticipants
also averaged over twice that of participants. Total assets, on
an adult equivalent basis, averaged $1,192 for participants and
$2,533 for nonparticipants. Those who are eligible, but do not
use stamps earn more money, receive less welfare, and have substantially greater assets than those who do receive stamps.

The greatest difference of all between the two groups was in
whether the household was headed by a man or a woman. Some 60
percent of participating households were headed by a woman,
compared to only 22 percent of eligible nonparticipating families, a pattern similar to the national one. In terms of
ethnic background, chicanos were somewhat less likely to use the
program, when eligible, than blacks or whites. Some 14.2 percent of the eligible nonparticipants and 8.7 percent of participants were chicanos.

EFFECTS OF FOOD STAMP PROGRAM ON FOOD EXPENDITURES
The effects of the FSP on food expenditures can be examined
in two alternative ways.[2] One approach is to compare expenditure
patterns of households who participate in the program to those
who do not participate. The second approach is aggregative and
identifies the effects of the program on the total demand for
food and subsequent impacts on food prices. Some results of
each of these approaches are summarized in the following sections. Other related studies are reported in USDA (1978).

Individual Household Expenditure Patterns
Comparisons of food expenditures by households participating
in the FSP to those who were not participating are available
from West's (1979) analysis of the 1973-74 BLS consumer expenditure data. His tabulations indicate average expenditures for

food at home by the 587 households participating in the FSP exceeded by approximately $1.30 per week the amount reported by nearly 10,000 households who were not participating. Substantially larger amounts, however, were spent for food away from home by nonparticipants relative to participants. Nonparticipants reported an average of $9.84 per week for food away from home compared to $2.31 for FSP participating households. This pattern is consistent with the fact that food stamps were to be used only to purchase food for at-home consumption. Also, many nonparticipating households had higher incomes than those participating in the FSP, undoubtedly causing differences in away-from-home food expenditures.

When comparisons were made on a per capita basis, FSP participants had lower expenditures for total food, food at home, and food away from home. Thus, the larger number of household members among FSP participating households was more than enough to offset the difference in household expenditures for food at home noted above. The difference in food-at-home expenditures was not statistically significant at the .05 probability level. Differences in total food and away-from-home food expenditures per capita between FSP participating and nonparticipating households were statistically significant.

A comparison of food expenditures was also made between FSP participating households and a group of low-income households who apparently were eligible but had chosen not to participate in the FSP. This comparison indicated the average at-home food expenditure per capita was nearly 10 percent larger for FSP participants than FSP eligible nonparticipants. On the other hand, total food expenditures per capita were slightly less for FSP participants relative to the values reported by eligible nonparticipants. Average expenditure per capita for food away from home by eligible nonparticipants was more than twice as large as the $1.01 per week average per capita for FSP participants. These data suggest that even though the difference in expenditures on total food was not statistically significant, the FSP may have resulted in a substitution between at-home and away-from-home expenditures. This interpretation is based on the strong, and perhaps unrealistic, assumption that the only difference between the two groups of households is participation in the FSP. Factors influencing whether FSP eligible households actually participate in the program may also affect food expenditures.

Additional analyses of how expenditures for at-home food were distributed among various product groups by FSP participants and FSP eligible nonparticipants revealed some significant differences. For example, FSP participants spent significantly more per capita than FSP eligible nonparticipants for pork, poultry, eggs, fresh whole milk, flour and other cereal products, processed vegetables, cabbage, carrots, corn, nonalcoholic beverages, and condiments. Snacks, yogurt, and ice cream were the only categories for which expenditures per capita by FSP elig-

ible nonparticipants were larger than FSP participants. Food
expenditures per capita for all of the remaining 29 individual
food categories considered were quite similar between the two
groups of households.

Other evidence concerning the effects of the FSP on household
food expenditure is provided by regression analysis of a sample
of FSP eligible households with 8- to 12-year-olds in 1971-73
from Washington by West et al. (1978). This study indicated
that receipt of food stamps increased the value of food consumed
by $5.14 per month, an increase of 13 percent over the average
for eligible nonparticipants. Receipt of free school lunches
had little effect on other food consumption among FSP participating households. On the other hand, eligible nonparticipants
reduced other food consumption by a corresponding amount when
free lunches were received. It is not clear whether receipt of
bonus stamps was the major reason for this difference in behavior
or whether the two groups just responded differently to the
receipt of free lunches.

Aggregate Analyses

According to the theoretical model used by Schrimper (1978),
the extent to which the FSP affects aggregate food demand depends
on two components. These are the share of the food market
accounted for by FSP participants and the effectiveness of the
transfer in food purchasing power transmitted by the program.

The first component can be crudely approximated by the
proportion of total population participating in the FSP. This
results in an overestimate of the market share since participants' per capita consumption would generally be less than the
average for the entire population. Consequently, the share of
the total food market accounted for by stamp recipients is likely to be somewhat less than their share of the total population.

The second component is a more elusive concept to quantify
since the total value of bonus stamps does not represent the net
increase in food expenditure. Purchase requirements were init-
ially established in an attempt to guarantee that some increase
in food expenditure would be achieved by the issuance of bonus
stamps. If an eligible household considered the purchase requirement to be too high, it would decide not to participate in
the program. On the other hand, households which previously had
been spending more for food than the purchase requirement would
find that bonus stamps increased specific purchasing power for
food as well as other products. An increase in general purchasing power would result because the participation in the program
enabled the household to reduce the proportion of its private
resources that was previously allocated to food in the absence
of the program. This represents an increase in the proportion
of the household's private resources which could be used to purchase other items as well as additional food. Thus, even if
households were to use all the food stamps they received for
food purchases, the net increase in food expenditures could be

less than the amount of bonus stamps since recipients would divert some of their income to other products. Since all households of the same size with similar incomes and other characteristics do not spend identical amounts for food, establishment of any specified set of purchase requirements would induce a particular distribution of participation and net increases in demand. Elimination of the purchase requirement as of 1 January 1979 meant stamp recipients have increased flexibility in using their income to maximize their satisfaction.

Several studies investigating the impact of bonus stamps under a purchase requirement suggest an increase of $0.30 to $0.65 in food expenditures for each dollar of bonus stamps. West's regression analysis of a combined sample of all FSP eligible households from the 1973-74 BLS Consumer Expenditure Survey produced an estimate of $0.54 as the extra food expenditures resulting from a dollar of bonus stamps. A lower coefficient of $0.38 was obtained using only those households who were actually participating in the FSP. The latter value is identical to the value based on California data reported by Lane (1978). A slightly lower value of $0.30 was estimated by West and Price (1976) from Washington data. The latter study reported that bonus stamps had a greater impact on food consumed among chicanos than among blacks or whites.

Some of the above estimates are a little lower than previous values reported by Reese et al. (1974). Their analysis indicated bonus stamps were between 50 to 65 percent effective in increasing food expenditures by participating households under the pre-1970 and later more liberalized purchase requirement. The results led them to conclude that bonus stamps were approximately twice as effective as comparable cash income supplements in expanding food expenditure for low-income families. Salathe (1980a) has argued however that estimates of the effect of bonus stamps may be misleading if possible discontinuities in the response function are not taken into account.

The above coefficients indicate the proportion of bonus stamps used to increase food expenditures rather than the rate of increase in food demand by FSP participants. Some simulations by Salathe (1980b) based on the BLS consumer expenditure data imply that the FSP may have increased demand for food among participants by 10 percent. Lane estimated an increase of approximately 11 percent in total value of food per person due to the FSP based on California data. Similarly, analysis of the Washington data implied an increase of 12 to 13 percent in food expenditures due to the FSP.

An estimate of the effective increase in food demand by FSP participants can be multiplied by the share of the food market attributable to the participants to produce a value of the relative change in total domestic demand for food. Thus if the FSP resulted in a 10 percent increase in demand for food by participants and they account for around 8 percent of the population, the increase in aggregate food demand would be 0.8 percent.

Price Effects

Any increase in aggregate demand for food will produce an increase in equilibrium price unless the aggregate supply of food is perfectly elastic in the short run and long run. Increases in demand resulting from the FSP can be combined with price elasticity estimates of retail demand and supply of food to calculate alternative effects of the FSP on food prices in Schrimper's model. For example, with -0.2 and 0.8 price elasticities of retail demand and supply, respectively, an 0.8 percent increase in total demand for food due to the existence of the FSP would produce an increase in food prices of 0.8 percent. The price changes for different elasticities of supply and market share were tabulated by Schrimper to illustrate potential bounds on price increases for each 10 percent increase in demand for food by FSP participants. An increase in the number of FSP participants would cause a greater increase in price. Increases in price would be different between the short run and long run as well as vary among individual products. Differences in price elasticities of supply and demand for alternative periods of adjustment would result in the initial price increases being larger than the ultimate price effect for any given increase in demand. The rate of increase in demand for different products would also vary with an increase in purchasing power for food. FSP participants would likely increase the demand for some products more than others. This is consistent with West's comparisons of at-home food expenditure on various products between FSP participants and FSP eligible nonparticipants.

The effect of bonus stamps on the prices of various food commodities has been empirically analyzed by Belongia (1979). Estimation of reduced form price equations using 1970-77 quarterly data produced positive coefficients for the variable representing the per capita value of bonus food stamps for all models, as expected. The results indicated that a 0.07 percent increase in the price of all food was associated with a 1 percent increase in the real value of bonus stamps per capita. Similarly, increases of 0.34 and 0.32 percent were found for the cereal and bakery products and meat groups. A much smaller insignificant coefficient resulted in the case of the price of dairy products. The latter result may be partly because the price of dairy products is more regulated and less responsive to market forces relative to other products.

IMPACT OF FOOD STAMP PROGRAM ON NUTRIENT INTAKE

Evaluation of the nutritional impact of food programs is difficult for a number of reasons.[3] First, the current dietary standards are incomplete and imprecise. There are roughly 45 essential nutrients, but Recommended Daily Allowances (RDAs) have not been established for many of these (Sexauer 1978). Second, people's nutritional needs vary by age, sex, body size,

activity, genetic makeup and physiological state. RDAs vary
only for age and sex differences. To account for the other differences, the RDAs are set at levels which assure that the nutritional needs of most healthy persons are met. They are set
above the average requirement and therefore groups who meet less
than 100 percent of their RDAs for certain nutrients may still
have adequate intakes. Third, current knowledge of the relationship between nutrient intake and health is sparse. The effects
of severe deficiencies of certain nutrients are well established
but the long-run consequences of either excesses or slight
deficiencies of specific nutrients are not known. In addition,
the effects of combinations of nutrients taken either in excess
or below recommended levels has just recently received the
attention of nutritionists.

The major sources of nutritional status information are
dietary intake, biochemical tests of blood and urine, clinical
examinations by doctors and dentists, and anthropometric
measurements. There are numerous problems with the use of
dietary intake for the evaluation of nutritional status. The
first is the problem of what percentage of the RDAs constitutes
adequate intake. Intake data is usually gathered by the use of
the 24-hour recall. The recall method is subject to error in
memory of the respondent as well as error in measuring quantities. Respondent bias in relating types of foods consumed is
also possible. Additionally, the same type of food may differ
in nutrient content. Growing conditions, processing, preserving,
storing, and cooking in the home affect nutrient content of food.

Many of the biochemical tests measure the effects of long-
run deficiencies in nutrient intake. They are, however, costly
to perform and subjects are generally more reluctant to participate than with the 24-hour recall. Other methods have problems
with standards as well as with the higher cost for professional
services.

Evaluations of the food stamp program have centered on the
question of whether or not the program increases nutrient intake
of the participants. These have not generally addressed the
question of whether or not increases in nutrient intake are
needed. The validity of the research rests on the proposition
that if the food stamp program does increase nutrient intake, it
can be used as a tool for improving nutritional well-being among
poorly nourished populations.

The question of nonnutritional benefits to recipients has
not been investigated. Questions regarding the degree to which
the program increases the palatability of the diet and general
psychological well-being of recipients have been ignored. Even
the degree to which the program alleviates temporary hunger has
not been researched.

The Hypothesized Relationship
Consider the typical hypothesized income-food expenditure
relationship. At low levels of income it has a positive slope

which decreases as income increases. When expenditures are
replaced by a food quantity index, the curve flattens out at a
lower income level. This is due to the substitution of more
expensive food. If the dependent variable is nutrient intake,
one would expect a relationship with income to be similar to
that with quantity of food. Recent estimates of income elasticities with food expenditures have been relatively low. For
example, West and Price (1976) estimated a value of 0.04 with
1972-73 Washington State data. One would expect a lower elasticity for the quantity of food and for nutrient intake. The
Washington State data included only households with school-age
children. Also, there were few households with extremely low-
incomes. Nearly all households with very little income from
other sources received Aid to Families with Dependent Children
(AFDC) payments. In comparison with other states these payments
were relatively high (Table 13.2).

TABLE 13.2. Public assistance: average monthly payments ($) to recipients, 1973

State	Old age assistance	Aid to families with dependent children (per family)	Aid to totally disabled
California	115	211	152
Pennsylvania	53	237	78
Washington	67	227	130
Alabama	73	74	79
Arkansas	68	112	82
Florida	83	107	93
Georgia	59	105	69
Kentucky	68	146	94
Louisiana	73	93	57
Mississippi	54	52	65
North Carolina	81	130	85
South Carolina	57	87	68
Tennessee	55	104	74
Texas	54	109	76
Virginia	82	172	102

Source: U.S. Bureau of the Census (1974, 1975).

Adrian and Daniel (1976) showed that nutrient consumption is
not highly responsive to income. Therefore, the effectiveness
of income transfers to alleviate nutritional deficiency may be
quite limited. A similar conclusion is also reached by Prato
and Bagali (1976), who assert that the relative economic efficiency with which additional food nutrients are acquired
decreases with respect to food expenditure. They argue that
although the FSP may increase the participant households' food
expenditure, it does not necessarily result in a nutritionally
better diet because the additional income may be spent on foods
having high nonnutritional components or on higher cost foods.

If the Food Stamp Program was merely a cash subsidy with no
restrictions on how the cash is to be used, the above considerations would lead one to suspect little, if any, effect on nutrient intake. However, the provisions of the program specify that
the stamps must be spent on food. A lower limit is placed on
the amount recipients spend for food. This restriction has led

to higher propensities to consume from bonus stamps than from current income. Estimates of the increase in food expenditures from a dollar in bonus stamps range from $0.30 to $0.65 (see previous section by Schrimper).

The effect of bonus stamps on nutrient intake would be expected to be less than that on expenditures. Households can substitute less expensive items for more expensive items but keep nutrient intake constant. If households were obtaining an inadequate quantity of food, an increase in income should increase the quantity of food consumed and, consequently, nutrient intake. The low-income elasticities currently found with expenditures indicate this condition is not the case. If, however, the increase in expenditures through the food stamp program enable the household to increase the "nutritional quality" of the diet, there could be increases in the intake of certain nutrients. For example, if the household purchases more fruits and vegetables, intake of vitamins A and C could be improved with food stamp participation. Thus, the food stamp program may have positive effects on intakes of some nutrients.

The Food Stamp Program may be an effective instrument for increasing nutrient intake of households with very low-incomes with the effect diminishing as incomes rise. Presently in the United States there are numerous federal and state programs that in effect place a floor under income. Social Security, unemployment compensation, and welfare are the dominant programs that provide such a floor. Welfare payments vary considerably from state to state so that relatively low-income households still exist in certain parts of the country (Table 13.2). In addition, other food programs such as the National School Lunch Program, the National School Breakfast Program, and the supplemental food program for women, infants and children (WIC) provide food for the economically disadvantaged at no cost. It is possible, therefore, that the food stamp program would raise nutrient intake substantially in the absence of other food and income supplement programs, but its effect may be small when considered an addition to other programs.

Review of Previous Studies

All studies reviewed were made before the spring of 1979 and thus will pertain to the provisions of the food stamp program prior to elimination of the purchase requirement (EPR). The 1977-78 Household Food Consumption Survey of the USDA which includes data taken during 1979 and 1980 on low-income households provides the first opportunity to analyze the program effects after EPR.

An evaluation of the initial food stamp pilot program was made by the USDA (1962). It was based on dietary surveys of participating and nonparticipating households in Detroit, Michigan, and in Fayette County, Pennsylvania. Significant increases in the percentages of households having a good diet

occurred in both localities. A good diet was defined as supply-
ing 100 percent or more of the RDA for each of 8 nutrients (USDA
1962).

The degree of success of the pilot project has not been
found in studies of the actual program. The following reasons
have been put forth (USDA 1962):

1. An experimental program is likely to receive a greater
   degree of cooperation, interest, and support than might
   prevail in the longer run.

2. USDA personnel maintained close supervision.

3. The families paid the amount they normally spent on food
   to receive their allotment of coupons.

4. Nutrition education was an integral part of the program.

Previous studies of the actual program have been limited to
a study by Madden and Yoder (1972) involving two rural central
Pennsylvania counties and a study by Lane (1978) in Kern County,
California.

The major conclusion of the Madden and Yoder study with
respect to the nutritional effects of the food stamp program was
that "food stamps provided some improvements in the diets of
families experiencing temporary shortages of funds--that is,
more than two weeks since payday." Lane concluded that "the
value of food available does appear to affect some nutrients,
but the effect is small." Thus, the evidence of these two studies suggests limited improvement in the intake of certain nutrients from food stamp participation. The improvement is not
universal and varies by time and place.

Results from Washington State Data

During the period 1972-73, a study of approximately 1,000
school children ages 8-12 was made in the state of Washington.
Dietary data were collected from the child at school with three
24-hour recalls. These were taken during different days of the
week to ensure obtaining weekend records and at different weeks
of the month to ensure data collection at different times during
the pay period. Household income and food stamp participation
data were collected with the use of an interview of the person
in charge of food preparation.

Food stamp participation and the household's food expenditures were included in models explaining nutrient intake of the
child (Price et al. 1978). Neither variable was significant at
the 10 percent level. These models included either liquid
assets or total assets which were significant for 4 of the 10
nutrients. These asset variables are correlated with food
expenditures and food stamp participation which leaves causal

relationships somewhat in doubt. The results of this study indicate that assets affect nutrient intake rather than food expenditures or food stamp participation. Since liquid assets can be used to purchase food in times of cash shortages, or since households with liquid assets are not likely to have cash shortages, these results tend to agree with those of Madden and Yoder.

A comparison of mean intakes does show small increases in nutrient intake from both food stamp participation and from higher expenditures for food (Table 13.3). The sharpest increases from food stamp participation were for riboflavin, vitamin A, and calcium. The increase in riboflavin intake was significant at the 10 percent level for the white sample. Since calorie intake was almost identical between the two groups, there is some indication of an increase in the nutrient quality of the diet (defined as the ratio of the intake of selected nutrients to calorie intake) with food stamp participation. Similar results are shown for different levels of household food expenditures (Table 13.3). This comparison showed the sharpest increases to occur with vitamin A and thiamin with riboflavin and calcium ranking third and fourth in the amount of increase.

TABLE 13.3. Percent RDA nutrient intake of 8 to 12-year-old children from households eligible for food stamps, state of Washington, 1972-73

Nutrient	Eligible non-recipients[a]	Food stamp recipients[a]	Percent change	Food expenditures greater than $44/mo/AE[b]	Food expenditures less or equal to $44/mo/AE[b]	Percent change
Energy	78.7	79.2	0.6	77.2	78.5	1.7
Protein	175.1	179.5	2.5	170.8	173.3	1.5
Calcium	94.6	101.2	7.0	88.9	94.8	6.6
Phosphorus	125.5	131.5	4.8	120.2	125.6	4.5
Iron	95.1	100.1	5.3	92.4	97.1	5.1
Vitamin A	122.0	131.5	7.8	115.1	128.0	11.2
Thiamin	105.9	109.1	3.0	95.7	105.1	9.8
Riboflavin	150.4	167.3	11.2	143.1	153.3	7.1
Niacin	93.9	98.9	5.3	89.8	94.4	5.1
Vitamin C	176.2	177.6	0.8	170.6	174.7	2.4

[a]Sample values were weighted by the state's proportion of blacks and Chicanos. This has the effect of keeping ethnic group constant.
[b]These are monthly food expenditures on an adult equivalent (AE) basis for the sample eligible for food stamps.

The comparison of means obtained from cross-section data is not a good method of evaluating the effect of food stamp participation. Relevant variables are not held constant. Food stamp recipients may have characteristics that differ from eligible nonrecipients that are related to nutrient intake. In the Washington State study it was found that food stamp recipients have a higher level of physiological need than do eligible nonrecipients. Additionally, household food expenditures were positively related to physiological need (West et al. 1978). Thus, without the program, those households who had received food stamps would have a higher food expenditure than those who did not.

SELECTED FOOD NUTRIENTS AVAILABLE TO LOW-INCOME
HOUSEHOLDS IN SOUTHERN REGION OF UNITED STATES

Although previous studies suggest that the FSP affects participant household's food expenditures, the evidence suggests
that the nutritional impact of the FSP is less than desired for
two reasons.[4] First, the magnitudes of nutrient consumption re-
sponsiveness to income are relatively small (Adrian and Daniel
1976). Thus, the ability of transfer income derived from the
FSP to increase the intake of essential nutrients is limited.
Second, food stamp participants may purchase more of the foods
they are accustomed to rather than items which remove nutritional deficiencies from their diet (Prato and Bagali 1976).

Using the state of Washington data, Price reported in the
previous section that little relationship existed between the
receipt of stamps and nutrient intake of school children. A
possible reason for the lack of a relationship is that Washington
State has considerably higher welfare payments than most states,
particularly those in the South. For example, the average monthly AFDC payments to low-income households was as much as four
times greater in Washington than in Mississippi in 1973 (Table
13.2). In general, low-income households in Washington received
about twice as much in AFDC payments per household as those in
the South. Therefore, it is possible that the FSP may have a
more positive nutritional impact on low-income households in the
South than suggested by Price using Washington State data.

In their study, Scearce and Jensen (1979) computed the
selected food nutrients available to low-income households in
the South from the 1972-73 BLS Consumer Expenditure Diary Survey.
The sample consisted of 216 food stamp households and 1,144
eligible nonparticipant households. Their objective was to
determine whether the FSP has a significant impact on availability of selected nutrients for low-income households in the
South, since nutritional information was derived from the BLS
food expenditure data. The procedure involves the selection of
70 food items to be used as a representative market basket of
foods purchased for at-home consumption.

Furthermore, the authors assumed that regional, monthly,
average retail prices of the selected food items could provide a
reasonable proxy for the price paid by the families in the
sample. Regional average prices were calculated from monthly
BLS reports. These prices were then used to convert the appropriate weekly expenditures into weekly quantity. To derive
amount of nutrients available to each household, the quantity of
each food item purchased was multiplied by the amount of nutrients in the food item and summed over all 70 foods. The nutritive value for a given quantity of food was obtained from Composition of Foods: Raw, Processed, Prepared (Watt and Merrill
1963).

On the strength of the sample evidence, Scearce and Jensen
conclude that the FSP had significantly increased the amount of

6 of the 9 essential nutrients available to participating low-income households in the South.

Given that low-income households in the southern region of the United States generally received considerably lower public assistance payments in addition to the FSP relative to those in other regions of the United States, the relatively greater effectiveness of the FSP reported in Scearce and Jensen's study seem reasonable. However, their model may be modified to incorporate the following considerations. Food stamps are in-kind transfer income designed to provide assistance to eligible low-income households to purchase food through normal marketing channels. Although income is one of the criteria used to determine the eligibility of a household in receiving food stamps, the maximum amount of food stamp allotment available to the eligible household is determined by the size of household. Thus, it would be logical to hypothesize that the FSP participant and eligible nonparticipant households respond to income change differently. A lower income coefficient for food stamp recipients than eligible nonrecipients would be expected. It is also desirable to ascertain the effectiveness of the FSP between participant and nonparticipant households at different stages of the family life cycle. Additionally, it would be more appropriate to account for variations in household age-sex composition than simply to account for variations in household size. In the next section, results from an alternative model specification are presented to provide additional empirical evidence on the impact of the FSP on the amount of selected nutrients available to low-income households for at-home consumption in the South.

Results and Discussion

The effects of the FSP on 9 selected nutrients including food energy, protein, calcium, iron, vitamin A, thiamin, riboflavin, niacin, and vitamin C were estimated separately using regression analysis (Huang and Scearce 1982).

Results of the regression analysis suggest that the effects of household income on 9 essential nutrients are all positive and significantly different from zero at the .05 significance level except for riboflavin. Furthermore, the signs of the interaction term between household income and participation in the FSP are all negative, as expected, suggesting that FSP participant households have lower income responses than eligible nonparticipating households. Statistical tests suggest that the null hypothesis (the effect of income on a selected nutrient is the same between FSP participant and nonparticipant households) could not be rejected for calcium and riboflavin at the .10 significance level. These results conform with what might be expected that, other things being equal, FSP participant households have lower income responses on the purchase of selected nutrients than eligible nonparticipant households.

Elasticities with respect to household income for FSP partic-

ipant and nonparticipant households, respectively, were computed
at mean values and presented in Table 13.4. The magnitudes of
the income elasticities of selected nutrients for FSP participant households are all very close to zero except for vitamin
A. This result suggests that for a given increase in household
income, other things being equal, FSP participant households
would not purchase additional food in excess of those available
from food stamps. Not surprisingly, this finding coincides with
what might be expected since the provisions of the FSP specify
that food stamps must be spent on food purchases. Thus, by
nature of the design of the FSP, a lower limit is placed on the
amount that FSP participants spend on food. Nevertheless, the
results also seem to imply that the FSP may have in effect
become an upper limit on the amount participants spend on food
purchases for at-home consumption. The results appear to suggest
that available nutrients derived from participation in the FSP
may not have responded to income changes at all. Additionally,
this finding may be attributed to the possibility that low-income
households may have a relatively higher degree of affinity for
nonfood items. Thus, participant households may choose to purchase more of other nonfood items instead of purchasing additional food for at-home consumption.

TABLE 13.4. Income elasticities for selected nutrients, FSP participant and
eligible nonparticipant households, Southern region United States, 1973–74

Nutrient	FSP participant households	FSP eligible nonparticipant households
Food energy	−.01	.23
Protein	.06	.18
Calcium	.01	.14
Iron	.03	.23
Vitamin A	.11	.27
Thiamin	−.03	.20
Riboflavin	−.03	.51
Niacin	−.07	.30
Vitamin C	.05	.22

The estimated income elasticities of selected nutrients for
nonparticipant households, as shown in Table 13.4, are also
quite inelastic. Those income elasticities suggest that the
amount of nutrients available to nonparticipant households are
not highly responsive to income changes either. The results are
in accord with the finding of previous studies which generally
conclude that income effect on nutrient consumption is quite
limited (Adrian and Daniel 1976; West and Price 1976). One
should note that the magnitude of the estimated income elasticity for each selected nutrient for nonparticipant households is
much smaller than the income elasticity reported by Scearce and
Jensen with the exception of thiamin, riboflavin, and niacin.
The effect of stages of household's life cycle on selected
nutrients was represented by the age of household head. Stage 1
of the family life cycle represents those households with the
age of the head less than 25 years. Stage 2 represents those

households with the age of the head 25 or older but less than
45. Stage 3 is defined as those households whose head is at
least 45 but not older than 65. Finally, if age of household
head is greater than 65, the household is in stage 4 of the
family life cycle.

The results indicate that the amount of most selected nutrients available for low-income households with head of household
less than 45 are consistently significantly less than those
households in which the household head was 45 or older. The
differences between life cycle stages 3 and 4, in most cases
however, were not statistically significant at the .10 significance level. Households in stage 4 of the life cycle have significantly greater amounts of calcium, vitamin A, and vitamin C
available for consumption than households in stage 3 of the life
cycle.

More significantly, the impact of the FSP on selected nutrients available to low-income households in the South is examined
within a given stage of life cycle. The results indicate a
positive effect but not statistically significantly different
between FSP participant and nonparticipant households except for
households in stage 4 of the life cycle. In contrast, it appears
that participation in the FSP, holding all other factors constant, has a significant negative effect on the availabilities of
all nutrients for households in stage 4 of the life cycle. This
result appears consistent with Senauer's finding that elderly
households had lower participation rates in the FSP. One possible explanation is that most of these households are elderly,
single households which use food stamps to substitute for their
regular income in food purchasing instead of using stamps to
increase their food purchases. Therefore, it is possible that
some eligible households in stage 4 of the family cycle spend
the same amount to purchase foods for at-home consumption after
participation in the FSP as they did prior to their participation in the program. The results seem to suggest that participation in the FSP allows those households in stage 4 of the life
cycle to purchase adequate amount of food for at-home consumption, therefore, they have no need to purchase additional food
in excess of those available from food stamps.

The degree of urbanization generally shows no significant
effects on selected nutrients available to low-income households
in the South. However, the results suggest that the amount of
selected nutrients available to black households were consistently less than nonblack households. Black households have significantly lower amounts of calcium, vitamin A, and riboflavin
available for consumption than nonblack households.

The female homemaker, if present, is usually responsible for
food purchasing. A higher degree of educational attainment by
the homemaker generally suggests an ability to relate food purchases to nutrients available for the household. Except for
iron, households with female homemakers who completed at least
high school education have greater amounts of nutrients avail-

able for consumption as compared to those households in which the homemakers had a lower level of educational attainment.

Limitations of the Study

One should be cautious about drawing policy implications from this study with regard to the effect of the FSP on diet and nutritional status of low-income households. Since the quantity of an individual food item purchased was derived from expenditures on the food item and average price, this may raise the question concerning the quality of the food items purchased. Food expenditure data reflect both the quantity and the quality of the food items. It is, therefore, possible that households may spend about the same amount on food while substituting quality for quantity of food purchased and, hence, do not actually increase availability of nutrients for consumption. Conceivably, the FSP may have resulted in providing participants with additional transfer income to upgrade the quality of their food purchases rather than increasing the amount of food purchased. If this is true, then the imputed selected nutrients available to FSP participant households may be overestimated and, consequently, the effect of the FSP on nutrients available to the households for at-home consumption may be positively biased toward a favorable conclusion.

Furthermore, the results are based on food nutrients available for at-home consumption, not individual dietary intake. Schrimper has pointed out in the previous section that households which are eligible but not participating in the FSP spent twice as much for food away from home than FSP participants. He suggested that the FSP may have resulted in a substitution between at-home and away-from-home food expenditures. Although away-from-home food expenditures generally do not account for any significant proportion of the total food expenditures among low-income households, the conclusion of the present analysis is only valid within the context of nutrients available to the households for at-home consumption. Further research is needed in order to assess the overall effectiveness of the FSP on nutritional status and dietary improvement of low-income households.

SUMMARY

Some specific questions related to the FSP were examined in this chapter. Based on both county data and household data, the first part of this chapter identified and provided insights into the factors that explain FSP participation. The analyses provide policymakers and program administrators with useful information which may help the FSP operate more efficiently and effectively. This information enhances the ability of policymakers and program administrators to predict program participation, given changes in the characteristics of the population, due to a factor such as recession. Just as importantly, the participation analyses identify groups such as the elderly which show lower participation rates. These analyses need to be carried out periodically, particularly following major policy changes such as elimination of the purchase requirement.

The effects of the FSP on food expenditures among participant and nonparticipant households and the aggregate impact of the FSP on the total demand for food and on food prices were examined in the second part of this chapter. Comparisons of food expenditures between households participating in the FSP and those not participating indicate differences in expenditure patterns. The nature of the differences depend on whether comparisons are made on a household or per capita basis for total, at-home, or away-from-home food expenditures. In 1973-74, FSP participating households reported greater expenditure per capita for food at home but less on food away from home and total food than eligible nonparticipants. Most comparisons of this type, however, involve problems associated with comparability of participant and nonparticipants in the FSP. An expansion in aggregate demand for food from the FSP involving a fraction of all U.S. households would tend to have a small positive effect on the retail price of food under a wide range of price elasticities of supply and demand. An effectiveness rate of 30 to 65 percent for bonus stamps with a purchase requirement has been estimated by several studies, implying a smaller increase in aggregate food expenditures than the value of bonus stamps. Empirical estimates of the effect of the FSP on prices of various food products tend to confirm theoretical implications.

Evaluations of the FSP have primarily concerned whether or not the program increases nutrient intake of the participants. The results of a survey of approximately 1,000 school children ages 8-12 conducted in the state of Washington are discussed in the third section. Although the increase in the intake of nutrients from participation in the FSP is, in general, not statistically significant, the Washington study suggests that there is some indication of an increase in the nutrient quality of the diet with FSP participation. The effect of multiple programs may be present. First, over 70 percent of the children from food stamp households were full participants in the school lunch program. Sixty percent of the food stamp eligible nonparticipants were full participants in the lunch program. Additionally, relatively high AFDC payments provided a relatively high floor under income.

Results from biochemical analysis along with the relatively high percentage of RDAs indicated very few problems of underconsumption of nutrients exist with the Washington State 8- to 12-year-olds. Thus if the food stamp program increased intake of certain nutrients, such an increase was not needed nutritionally. The most serious nutritional problem identified was obesity. Present food programs are not designed to deal with this problem.

The impacts of the FSP on selected nutrients available to

low-income households in the South are examined in the final section of this chapter. Based on selected nutrients calculated from the 1972-73 BLS Consumer Expenditure Diary Survey, the analysis suggests some significant differences in selected nutrients obtained for at-home consumption among low-income households in the South. The increase in availability of most essential nutrients for FSP participants provides empirical evidence to support the hypothesis that the FSP can positively and significantly increase the amount of food nutrients available for at-home consumption to low-income households in the South. The FSP has apparently failed to attain its desired effectiveness among those households whose head is 65 years old or older. This result may be attributed to lower participation rates among the elderly households. In the South, where the poverty rate is the highest and the public assistance payments are the lowest among all regions, the FSP aid to the eligible low-income households may be expected to be more effective than that of other regions in the United States.

NOTES

1. Author of this section is Benjamin H. Senauer.
2. Author of this section is Ronald A. Schrimper.
3. Author of this section is David W. Price
4. Author of this section is Chung L. Huang.

REFERENCES

Adrian, J., and R. Daniel. 1976. Impact of socioeconomic factors on consumption of selected food nutrients in the United States. Am. J. Agric. Econ. 58:31-38.

Belongia, M. 1979. Domestic food programs and their related impacts on retail food prices. Am. J. Agric. Econ. 61:358-62.

Hines, F. 1975. Factors related to participation in the food stamp program. USDA, ERS, Agric. Econ. Rep. No. 298.

Huang, C. L., and W. K. Scearce. 1982. Impact of the food stamp program on selected food nutrients purchased by low income households. Socio-Econ. Plann. Sci. 16:1-7.

Hunter, C. A., Jr. 1980. The demand for food stamps in North Carolina: A cross county analysis. Thesis, North Carolina State Univ.

Lane, S. 1978. Food distribution and food stamp program effects on food consumption and nutritional "Achievement" of low income persons in Kern County, California. Am. J. Agric. Econ. 60:108-16.

Madden, J. P., and M. D. Yoder. 1972. Program evaluation: Food stamp and commodity distribution in rural areas of central Pennsylvania. Pa. Agric. Exp. Stn. Res. Bull. No. 780.

Prato, A. A., and J. N. Bagali. 1976. Nutrition and nonnutrition components of demand for food items. Am. J. Agric. Econ. 58:563-67.

Price, D. W., D. A. West, G. E. Scheier, and D. Z. Price. 1978. Food delivery programs and other factors affecting nutrient intake of children. Am. J. Agric. Econ. 60:609-18.

Reese, R. B., J. G. Feaster, and G. B. Perkins. 1974. Bonus food stamps and cash income supplements: Their effectiveness in expanding demand for food. USDA, ERS, Mark. Res. Rep. No. 1034.

Salathe, L. E. 1980a. Food stamp program impacts on household food purchases: Theoretical considerations. Agric. Econ. Res., 32:36-40.

_____. 1980b. Impact of elimination of the food stamp program's purchase requirement on participants' food purchases. South. J. Agric. Econ. 12:87-92.

Scearce, W. K., and R. B. Jensen. 1979. Food stamp program effects on availability of food nutrients for low income families in the southern region of the United States. South. J. Agric. Econ. 11:113-20.

Schrimper, R. A. 1978. Food programs and the retail price of food. Agricultural-food policy review: Proceedings of five food policy seminars. USDA, ESCS-AFPR-2:101-8.

Sexauer, B. 1978. Food programs and nutritional intake: What evidence? Agricultural-food policy review: Proceedings of five food policy seminars. USDA, ESCS-AFPR-2:39-43.

Sexauer, B., R. Blank and H. Kinnucan. 1976. Participation in Minnesota's food stamp program. Minn. Agric. Econ. 576:1-8.

U.S. Bureau of the Census. Statistical abstract of the United States. 1974, 1975. Washington, DC: U.S. Government Printing Office.

U.S. Department of Agriculture. 1962. The food stamp program: An initial evaluation of the pilot projects. AMS-472.

_____. 1977. Agricultural statistics 1977. Washington, DC: U.S. Government Printing Office.

_____. 1978. The food stamp program: A review of selected economic studies. ESCS-34. 1978.

_____. 1979. Agricultural statistics 1979. Washington, DC: U.S. Government Printing Office.

Watt, B. K., and A. M. Merrill. 1963. Composition of foods: Raw, processed, prepared. USDA, ARS, Agric. Handb. 8, Rev. 1963. Washington, DC: U.S. Government Printing Office.

West, D. A. 1979. Effects of the food stamp program on food expenditure: An analysis of the BLS consumer expenditure survey 1973-74 diary data. Rep. to Food and Nutr. Serv., USDA.

_____, and D. W. Price. 1976. The effects of income assets, food programs, and household size on food consumption. Am. J. Agric. Econ. 58: 725-30.

West, D. A., D. W. Price, and D. Z. Price. 1978. Impacts of the food stamp program on value of food consumed and nutrient intake among Washington households with 8-12 year old children. West. J. Agric. Econ. 3:131-44.

Index

Adding-up (Engel aggregation), 6
    29, 94, 95, 103, 144, 156.
    See also Demand functions,
    properties
Additivity, 11-13, 15, 16, 17
  block, 15
  direct, 11-12
Aggregation, 26-27, 65-68, 172.
    See also Demand, key issues
Antisymmetry. See Preferences,
    axioms
Autocorrelation
  specification, 118
  tests, 121-23

Beckerian approach. See Budget
    constraint; Utility func-
    tion; Utility maximization
Block additivity. See Additivity
Budget
  constraint (income), 5, 7,
    20-22, 144
    Beckerian, 22
    Lancastern, 20
    multiperiod, 21
  forecasting household. See
    Linear logit model
  share
    average, 7, 93, 99, 138, 144
    marginal, 24, 92, 100, 119

CES. See Household survey data,
    Consumer Expenditure
    Surveys

CGCM. See Hedonic index model,
    consumer goods character-
    istics model
Comparability. See Preferences,
    axioms
Complement good, 9, 93, 131, 137
Concentration curve, 207-10
  defined, 207
  elasticity, 207, 209-10
  functional forms, 207-8
Consumer expenditure surveys,
    problems in, 62-82
  data problems, nature of, 63-
    74
  effects, on resulting estimates,
    74-78
  modeling problems, 78-82
Continuity. See Preferences,
    axioms
Convexity, 4, 6. See also
    Preferences, axioms
Cost function, 8
Cost-of-living index, 187, 206
Cournot aggregation, 6, 7, 29,
    144. See also Demand
    functions, properties
CREST. See Household panel
    data, Chain Restaurant
    Eating-out Share Trend

Data, 33-38
  accuracy, 34-36
  aggregative vs. cross-
    sectional, 36-37
  definitions, 38

market levels, 37–38. See
    also Demand, key issues
price linkage, 36
problems, 54–84
consumer expenditure survey.
    See Consumer expenditure
    surveys, problems in
disappearance data. See
    Disappearance data,
    problems in
selection of, 35
sources, public vs. private,
    33–34. See also Data
    systems
systems. See Data systems
time frame, 36
Data systems, 38–51
channel movement, 48–49
disappearance, 39, 183
    price, 39–41
economic censuses, 49, 97
household panels, 41–45
household surveys, 46–48
new systems, 50–51
    attitudinal and innovation
        adoption, 51
    continuing household survey,
        50
    retail scanner, 50–51
store panels, 45–46
Demand
functions
    derivation, 5–6
    functional forms, 23, 24–25,
        78–80, 91, 172
    Hicksian, 8
    Linear expenditure system,
        92. See also Linear
        expenditure system
    Marshallian, 6
    multiperiod, 21
    properties, 6–10
    S_1-branch system, 130
key issues, 23–28
neoclassical theory, 3–17, 19,
    20, 25, 26
neoclassical theory, exten-
    sions of, 17–23
systems. See Demand systems
Demand systems
complete
    addilog, 25
    AIDS (almost ideal demand
        system), 25, 143
    Australian, 25

complete vs. partial, 23–
    26, 28, 171–72
constant elasticity, 25
dynamic, 26, 114–15
inverse, 157
linear expenditure, 25,
    91–95. See also
    Linear expenditure
    system
linear logit. See Linear
    logit model
multinomial logit, 25
policy analysis, applica-
    tion for, 157–60
Rotterdam, 25
static, 25, 114
S_1-branch. See S_1-branch
    system
translog, 25, 143
partial
complete vs. partial. See
    Demand systems,
    complete
disequilibrium hypothesis,
    test, 182–83
dynamic, 173–75
expenditure patterns, 172–73,
    178–80, 188–94
food stamp program analysis.
    See Food stamp program
Lorenz and concentration
    curves, analyses, 206–10
nutrients, demand for. See
    Hedonic index model;
    Nutrient intake
psychological and social
    variables in, 194,
    200–2
and quality of life, 202–6
regional demand, 180–82
retail and wholesale level
    demand, 176–78
socioeconomic and demo-
    graphic variables in,
    188, 191, 194–200
Differentiability, 4, 6. See
    also Preferences, axioms
Disappearance data, problems in,
    55–62
demand models, effect on,
    57–62
measurement error, 55–57
Divisibility. See Neoclassical
    theory, commodity set
Duality, 8

Elasticity
  age, 148, 205
  cross-price, 7, 15, 134-37,
    145
    compensated, 24, 93, 94, 96
    long-run vs. short-run, 174-
      75, 198
    uncompensated, 24, 103,
      104, 156
  expenditure, 102, 134-37,
    191-94
  formulas, for LES, 96
  household size, 134-37, 148
  income, 7, 8, 10, 15, 24, 145,
    156, 181, 198, 204-6,
    209-11, 272-73
  long-run vs. short-run, 26
  of marginal utility of income,
    8, 24
  matrix, 155, 156, 158
  own (direct) price, 7, 10,
    134-37, 145, 181
    compensated, 24, 96
    long-run vs. short-run, 26,
      174-75, 177
    uncompensated, 24, 103, 104,
      156
  of substitution, 24, 93, 94,
    102, 130
Engel
  aggregation. See Adding-up
  curve (function), 76, 80, 81,
    172, 178, 188-91, 203-4,
    207
Equivalent scales, 187, 189-91,
  195, 202-6

Flexibility matrix, 155, 157,
  158
  quantity change analysis, 157
  weighting procedure, 158
Food stamp program
  demand and price, effect on,
    263-65
  expenditure, effect on, 261-
    63
  history, 255-57
  nutrient availability, effect
    on, 271-75
  nutrient intake, effect on,
    248, 265-70
  participation, factors
    affecting, 257-61
Frisch's money flexibility, 8

Full information maximum likelihood, 118, 119, 134

Giffen paradox (inferior good),
  12, 93, 131

Habits
  defined, 115-16
  linear, 117-18, 123
  proportional, 119-20, 123
  state adjustment, 116
  tests, 121-23
Hedonic index model
  consumer goods characteris-
    tics model (CGCM), 221
  estimation, 224-26
  implicit price, 221-23, 226-32
  specification, nutrient demand,
    222-24
Hessian matrix, 6, 8, 10, 11,
  12, 15
Hicksian approach. See also
    Budget constraint; Utility
    function; Utility maximiza-
    tion
Hicksian fundamental equation of
  value theory, 9
Homogeneity, 6, 7, 29, 94, 95,
  144, 156, 160, 178-80. See
  also Demand functions,
  properties
Household budgets, forecasting.
  See Linear logit model
Household panel data. See also
  Data systems
  Griffin panel, 41, 172, 178
  Market Research Corporation
    of America (MRCA), 42
  Michigan panel, 44-45
  National Purchase Diary
    Research (NPDR), 42-44
    Chain Restaurant Eating-out
      Share Trend (CREST),
      43-44
  Puerto Rico panel, 41-42
Household survey data. See also
  Data systems
  Consumer Expenditure Surveys
    (CES), 47-48, 132, 147,
    188, 194, 203, 208, 259,
    271
  household food consumption
    survey, 48, 188, 189

Income effect, 9
Independent good, 9
Inequality coefficients, 207–9
Inferior good. See Giffen
    paradox
Klein–Rubin utility function. See
    Utility function

Lagrangian multiplier, 5, 6
Lancastern approach, 19–20. See
    also Budget constraint;
    Utility function; Utility
    maximization
Leontief–Hicks composite com-
    modity theorem, 10
Leser's approximation, 93–94
    See also Linear expenditure
    system
Likelihood ratio, 121–23
Linear expenditure system (LES)
    elasticities, 96, 102–5,
        120–21, 124
    estimation, 98–99, 119
    habits specification, 117–18,
        120
    Leser's approximation, 93–94
    persistence and autocorrela-
        tion, tests, 121–23
    Powell's approximation, 94–95
    Stone's formulation, 92–93
Linear logit model, 143–52
    elasticities, 145–46, 148
    food budgets, forecasting,
        149–52
    household size and composition
        effects, 146–49
    specification, 144–45
Lorenz curve. See Concentration
    curve, defined
Luxury good, 150

Marginal rate of substitution,
    14, 15–16
Marginal utility
    of commodity (good), 5, 11,
        15, 93
    diminishing, 6
    of income, 5, 6, 8, 9, 11, 24
Monotonicity, 4, 5, 6. See also
    Preferences, axioms
MRCA. See Household panel data,
    Market Research Corpora-
    tion of America

Neoclassical theory, 3–5, 171
    commodity set, 4
    extensions of, 17–23
        effects, 17–18
        multiperiod decisions, 20–21
        new commodities, 19–20
        risk and uncertainty, 19
        time, 22–23
    preference axioms, 4–5
    utility function, 4, 172
Nielson, A. C., data service.
    See Data systems, store
    panels
Nonnegativity. See Neoclassical
    theory, commodity set
NPDR. See Household panel data,
    National Purchase Diary
    Research
Nutrient intake
    assessment, 250–51
    children
        5- to 12-year-old, 237–42
        8- to 12-year-old, 244–50
    food stamp participation, 248
    school lunch participation,
        245–46
Pearson product–moment correla-
    tion coefficients, 138–39
Persistence. See Habits
Powell's approximation, 94–95.
    See also Linear expenditure
    system
Preferences
    additive, 11–12. See also
        Additivity
    almost additive, 11, 12–13,
        16, 17
    axioms, 4
    ordering, 4, 7
    separability, 11, 14–17
        Pearce, 14, 15–17, 130
        strong, 14, 15, 16, 17, 93,
            95
        weak, 14–15, 16, 17
    structure, 10–17
Price change
    analysis of, 156
    compensated, 8
    marginal-utility-of-income-
        compensated, 9, 10
    uncompensated, 7, 9

Quasi-concave. See Preferences,
    axioms

RDA. See Recommended Daily
    Allowance
Recommended Daily Allowance
    (RDA), 223, 225, 237, 265
Roy's identity, 8

SAMI. See Data systems, channel
    movement
S₁-branch system, 128–41
  branches, defined, 132
  demand function, 130
  elasticities, 134–37
  estimation, 134
  goodness-of-fit, 138
  predictive performance, 138
  sociodemographics, incorporat-
    ing, 131
Separability. See Preferences
Shephard's Lemma, 8
Slutsky symmetry, 6, 7, 9, 29,
    94, 144, 156, 160. See also
    Demand functions, properties
Stone-Geary utility function.
    See Utility function
Stone's formulation, 92–93.
    See also Linear expenditure
    system
Strotz-Gorman utility tree, 129–30
Substitute good, 9, 131, 137
Substitution effect
  cross, 10, 95, 150
    direct, 10, 150
  general, 9
  Slutsky, 93–95
  specific, 9, 10, 13
Supernumerary income, 93, 130

Transitivity. See Preferences,
    axioms

Unboundedness. See Neoclassical
    theory, commodity set
Utility function. See also
    Preferences, separability
  Beckerian, 22
  direct, 7, 8, 18, 25, 129
  discounted, 26, 117
  dynamic, 26
  indirect, 7, 8, 18, 25
  Klein-Rubin (Stone-Geary),
    11, 92, 100
  Lancastern, 20, 220–21
  Leontief, 188
  multiperiod, 21
  quadratic, 10
Utility index
  complete, 19
  constant, 8
  multiperiod, 21
Utility maximization
  Beckerian, 22
  first-order conditions, 5, 6
    multiperiod, 21
  Lancastern, 22
  second-order conditions, 6
    multiperiod, 21
  Strotz's two-stage procedure,
    14, 29

Vertical linkage, 27–28, 172.
    See also Demand, key issues

Welfare analysis, 8, 186–88, 206